T0304777

Voices of Innovation

Everyone talks innovation and we can all point to random examples of innovation inside of healthcare information technology, but few repeatable processes exist that make innovation more routine than happenstance. How do you create and sustain a culture of innovation? What are the best practices you can refine and embed as part of your organization's DNA? What are the potential outcomes for robust healthcare transformation when we get this innovation mystery solved? Through timely essays from leading experts, the first edition showcased the widely adopted healthcare innovation model from HIMSS and how providers could leverage to increase their velocity of digital transformation. Regardless of its promise, innovation has been slow in healthcare.

The second edition takes the critical lessons learned from the first edition, expands and refreshes the content as a result of changes in the industry and the world. For example, the pandemic really shifted things. Now providers are more ready and interested to innovate. In the past year alone, significant disruptors (such as access to digital health) have entered the provider space threatening the existence of many hospitals and practices. This has served as a giant wake-up call that healthcare has shifted. And finally, there is more emphasis today than before on the concept of patient and clinician experience. Perhaps hastened by the pandemic, the race is on for innovations that will help address clinician burnout while better engaging patients and families.

Loaded with numerous case studies and stories of successful innovation projects, this book helps the reader understand how to leverage innovation to help fulfill the promise of healthcare information technology in enabling superior business and clinical outcomes.

Voices of Innovation

Fulfilling the Promise of Information Technology in Healthcare

Second Edition
Edited by Edward W. Marx

Foreword by Dr. Tomislav Mihaljevic,
President and Chief Executive Officer,
Cleveland Clinic

Introduction by Jennifer Fox,
Chief Executive Officer,
Mary Crowley Cancer Research

Routledge
Taylor & Francis Group

A PRODUCTIVITY PRESS BOOK

Second edition published 2024
by Routledge
605 Third Avenue, New York, NY 10158

and by Routledge
4 Park Square, Milton Park, Abingdon, Oxon, OX14 4RN

Routledge is an imprint of the Taylor & Francis Group, an informa business

1st edition published by Routledge 2019

ISBN: 978-1-032-44527-4 (hbk)
ISBN: 978-1-032-44525-0 (pbk)
ISBN: 978-1-003-37260-8 (ebk)

DOI: 10.4324/9781003372608

Typeset in Garamond
by Apex CoVantage, LLP

Contents

Foreword

This is a dangerous book. It's a threat to complacency and the way things have always been done. It provides a direct challenge to all of us who work in healthcare—doctors, nurses, and administrators at all levels.

Why is this? Because, astonishingly, healthcare is the only major industry that has yet to be transformed by the digital revolution. While manufacturing, retail, and transportation enjoy gains in productivity from digital technology, some sectors of healthcare are still struggling to adopt the electronic medical record, telemedicine, and the host of other game-changing innovations barreling down the digital pipeline.

Voices of Innovation shows us that it doesn't have to be that way. A new generation is taking the helm, energized by the potential of IT to make healthcare safer, more effective, and more affordable.

At Cleveland Clinic, our entire organization is moving rapidly to adopt digital technology across all clinical domains. Our surgeons who performed the world's first 100% face transplant planned their stupendously complex procedure using patient-specific 3D-printed models assembled from CT and MRI data. We've partnered with top companies to build powerful databases and systems to enhance the speed and accuracy of clinical decision making. With a variety of hand-held apps, we're monitoring hundreds of patients at home, keeping them out of the hospital, and responding quickly to acute emergencies.

Almost every place evidenced-based digital solutions are applied, providers are seeing favorable clinical outcomes, improved patient satisfaction, and lower costs.

We're rapidly approaching the day when the majority of clinical decisions will be supported by information from wearable devices, imaging, implants, genetic profiles, and analysis of the microbiome—with insights from global health and all published research.

This is not a future where doctors are replaced by computers. It's a future where doctors have access to 1,000 times more knowledge than they could ever get from their own reading. It's a future where doctors are liberated from superficial tasks and can focus their full attention on the patient and the patient's problem.

The people you'll meet in *Voices of Innovation* are at the leading edge of all these trends. They aren't dreaming about the future of healthcare IT: they're making it happen right now. And reading about what they're doing will make you want to jump out of your seat and begin innovating yourself, whatever your sphere.

Digital technology is going to help us do what healthcare does well, even better. We will use it to connect people with people in new ways—to humanize care and enhance our role as healers.

Dr. Tomislav Mihaljevic,
President and Chief Executive Officer, the Cleveland Clinic

Introduction

I first met Edward W. Marx in 2022 as a member of the board of directors. As the CEO of Mary Crowley Cancer Research, I find it critical to engage board members who believe in innovation and transformation. It has been invigorating to add someone to our board of experts with broad healthcare experience who is a digital transformation and innovation expert. When we met, I knew we had the right level of expertise to join our board. Having previously served as the VP of Research Operations for a large healthcare organization in a major Texas market, I am familiar with the need to be relevant and keep up with the digital transformation that is possible today. I believe that innovation will enhance and transform the patient experience and elevate healthcare transformation to unprecedented levels. We are fortunate to have great innovators such as Ed to propel our organization and industry forward.

Ed has a deep history in healthcare when it comes to innovation. In fact, he has a reputation for collaborating with teams and developing innovative solutions to maximize technical and digital investments. Early in his career, he and his team developed the first two-way NICU video solution so parents could have 24x7 interactions with their sick babies. Later, Ed's teams would develop creative solutions that helped save lives and improve the quality of care at multiple organizations, including Texas Health Resources and Cleveland Clinic.

If you have followed Ed, you understand he does not sit on his laurels but continues to be a catalyst pushing the industry forward. As the chief digital officer for Tech Mahindra, his teams developed the world's first complete virtual care platform. In 2022, Ed became the chief executive officer of healthcare advisory and solutions firm Divurgent and has flipped the traditional consulting model further, differentiating their go to market strategy and offerings from traditional competitors. I can't wait to see what Ed does next!

Ed has also leveraged a variety of media tools to further encourage not just innovation but digital transformation. He is an avid blogger, and his essays have more than 100,000 visits. His podcast, DGTL Voices, is ranked in the top 3%. He has written several healthcare best-selling books plus a "coffee table" book on running and a collection of stories from his first 50 years, titled *Extraordinary Tales of a Rather Ordinary Guy*. He and his wife, Dr. Simran Marx, plan to release a book in 2023 on romance in marriage.

After the success of the 2019 healthcare bestseller *Voices of Innovation*, Ed shared he was working on a second edition and asked me to contribute. He also asked that I write this introduction. How could I say no? I understand firsthand the importance of innovation in all aspects of healthcare, from payer to bio tech to providers to research and clinical trials. As a Phase I clinical trial center, Mary Crowley Cancer Research success to date is all around innovations of all kinds. Cancer research innovation translates into saving more lives and creating hope.

What makes this book even more special is that 100% of all author royalties goes directly to Mary Crowley so we can continue to invest in finding new cancer treatments that are both more effective and less toxic to the patient. Hope lives here.

Jennifer Fox,
Chief Executive Officer, Mary Crowley Cancer Research

Acknowledgments

No book can be written alone. *Voices of Innovation* is no exception. There are many individuals and organizations who gave birth and life to this endeavor.

First, thank you to the fantastic organizations and teams I have had the honor to serve with. They taught me innovation and allowed me to fail and succeed along the way. It all begins with Poudre Valley Hospital in Northern Colorado, who allowed me to experiment and create unique programs and services for our medical staff, and Parkview Episcopal Medical in Southern Colorado, who gave me my first wings into healthcare technology. We learned that you could flip market share through the innovative application of technology. When I was caught in a long-term project pause at HCA in Tennessee, leadership allowed me to be one of the first in the world to experiment with personal digital assistants, which still informs my thoughts today. University Hospitals in Ohio provided me my first opportunity to collaborate with scientists and providers to co-create and develop repeatable processes along the way. Arlington-based Texas Health Resources was one of those "right place at the right time" scenarios where we were able to develop life-saving innovation on top of mature electronic health records, moving swiftly into mobile. With New York City Health & Hospitals, we applied innovation retrospectively in order to set the stage to leap frog competing institutions. Cleveland Clinic is the first organization I served where innovation is part of the DNA of the organization. Surrounded by an innovation superstructure and brilliant researchers, educators and providers, I learned all over again. Now as a consultant, it is the summation of all these experiences in a singular organization I am privileged to lead and serve. Within each of these organizations were mission-driven leaders and teams that inspired me and taught me things I could never get from a book, a speech or consultant. I get to help others innovate to improve care quality, reduce costs and, ultimately, save lives.

Second, my contributors. *Voices* is full of diverse practical examples of innovation in practice from around the globe. They each went through a lengthy period to have their work included, some as long as four years! As I read every submission, my eyes were opened and my mind enlightened, leaving me spinning full of new ideas.

Third, my support team has been patient and amazing, including editor, Kristine Mednansky at CRC Press/Taylor & Francis Group. She is a joy to work with and keeps me on task. I was thrilled when Kris asked me to write this second edition. HIMSS, who developed the initial idea and asked me to write it, have also been patient throughout this process and were helpful in the solicitation of contributors. The primary reason for the first edition of *Voices* to finally be published was the tenaciousness of my excellent assistant at the time, Dara Dressler. In addition to her work and in her spare time, Dara served as the project manager and ensured we stayed on track. *Voices* would not have been finished without her. For this second edition, *Voices* was enabled by my friend and editor Michelle Noteboom. She has always been my friend advocate and helped edit the first edition as well! A final shout out to my youngest daughter Shalani Shawn, who helped me process the second edition of *Voices*.

Finally, my Indian Princess and Wife, Dr. Simran Marx: Simran has provided continuous encouragement along the journey. Without complaint, she afforded me the time needed to complete *Voices*. Both editions! As a healthcare provider herself, Simran understood the importance of this work and the need to help others with all of the examples from our contributors. A goal of *Voices* is to make the practice of medicine easier and less burdensome for clinicians like Simran. She shares in the sacrifice and joy it took to write *Voices*.

Edward W. Marx is Chief Executive Officer of Marx & Marx, a global healthcare advisory and services firm. Prior to Marx & Marx, Edward served as the CIO of the Cleveland Clinic where he also led digital strategy. Prior to the Clinic, he served as Senior Vice President/CIO of Texas Health. In 2015, he spent more than two years as Executive Vice President of the Advisory Board, providing technology leadership and strategy for New York City Health and Hospitals. Edward began his career at Poudre Valley Health System. CIO roles have included Parkview Episcopal Medical Center, University Hospitals in Cleveland and Texas Health Resources. Concurrent with his healthcare career, he served 15 years in the Army Reserve, first as a combat medic and then as a combat engineer officer. Edward is a Fellow of the College of Healthcare Information Management Executives (CHIME) and Healthcare Information and Management Systems Society (HIMSS). He has won numerous awards, including HIMSS/CHIME 2013 CIO of the Year, and has been recognized by both CIO and Computer World as one of the "Top 100 Leaders." Becker's named Marx as the 2015 "Top Healthcare IT Executive" and the 2016 "17 Most Influential People in Healthcare." This is Edward's fifth book. His first book, *Extraordinary Tales of a Rather Ordinary Guy*, was published in 2014. The book was written in response to requests for him to detail the success principles behind many of his experiences. After writing the first edition of *Voices*, he published, *Scenes from an Early Morning Run*, in Spring 2019. It is a "coffee table" book that showcases unique photos he has taken while running over the last several years. This collage was collated in response to requests to have all these pictures in one volume. His last book was the 2020 healthcare best seller *Healthcare Digital Transformation: How Consumerism, Technology and Pandemic Are Accelerating the Future*. Edward has three other books in the making. He is now working on a first edition *Voices* for the payer community. He is collaborating with best friend and Mayo Clinic CIO, Cris Ross, on a yet to be titled book on patient and family experience. Finally, he and his wife, Dr. Simran Marx, have written a book on passion in marriage. Edward received his Bachelor of Science in psychology and a Master of Science in design, merchandising and consumer sciences from Colorado State University. Edward is married and proud father to five adult children and four grandchildren. In his spare time, he enjoys climbing some of the world's highest peaks and competing in multi-sport races as part of TeamUSA Duathlon.

About the Editor and Contributors

Rachini Ahmadi-Moosavi, MHA—Rachini is Vice President of Enterprise Analytics at UNC Health. She is passionate about empowering people, delivering value and improving lives. As a recent graduate of the Carnegie Mellon University CDataO Certificate program, Rachini continues to champion and lead data and analytics enablement within her Enterprise Analytics & Data Sciences team and across the analytics community at UNC Health, ensuring that healthcare transformation continues to be catalyzed by insights, interoperability and automation.

Pamela Arora—Pamela Arora brings expertise in healthcare technology, health facilities operations and community/industry boards. Her board leadership experience includes serving on the HIMSS North America Advisory Board, the Association for the Advancement of Medical Instrumentation (AAMI) Board, the Philips Pediatric & Neonatal Medical Advisory Board, the Health Information Trust Alliance (HITRUST) Board, the Tech Titans Advisory Board, the HIMSS Health Executive Institute Advisory Board and the College of Healthcare Information Management Executives Board. As President and CEO of the Association for the Advancement of Medical Instrumentation (AAMI), which advances the development and safe and effective use of medical technology, Pamela is responsible for championing a culture of openness and innovation, identifying and bringing together key constituencies to solve complex issues and balancing domestic and international points of view. She serves as AAMI's principal spokesperson, maintaining regular contact with member organizations, regulators, industry, media and the public, including relationships with the U.S. Food and Drug Administration (FDA), the Centers for Medicare & Medicaid Services (CMS) and other key regulatory agencies to ensure inclusion and further establish AAMI as an expert reference for those organizations.

This follows her time as Senior Vice President Strategic Technology of Children's Health in Dallas, TX, one of the largest pediatric non-profit health systems in the United States. Pamela also served as the Senior Vice President and Chief Information Officer, responsible for all information systems and technology, health information management and health technology management (biomedical technology) and support. Earlier, Pamela was Senior Vice President, Chief Information Officer at UMass Memorial Health Care, the largest healthcare system in Central and Western Massachusetts. She was the Interim Chief Executive Officer for LiquidAgents Healthcare. Before this, she held a number of positions of increasing responsibility at Perot Systems Corporation, culminating in Chief Information Officer. Prior to Perot, she delivered IT and process solutions for Electronic Data Systems (EDS) and General Motors Corporation. Pamela began her career as Programmer for Wayne State University.

Daniel Barchi—CommonSpirit CIO—former New York-Presbyterian—Daniel Barchi is former Senior Vice President and Chief Information Officer of New York—Presbyterian, one of the largest healthcare providers in the United States, as well as the university hospital of Columbia University and Cornell University. He leads 2,000 technology, pharmacy, informatics, artificial intelligence and telemedicine specialists who deliver the tools, data and medicine that physicians and nurses use to deliver acute care and manage population health. Daniel previously led healthcare technology as Chief Information Officer of Yale New Haven Health System and the Yale University School of Medicine, and earlier as Chief Information Officer of the Carilion Health System. Before those roles, Daniel was President of the Carilion Biomedical Institute and Director of Technology for MCI WorldCom. Daniel graduated from Annapolis where he served as Brigade Commander, began his career as a United States Naval officer at sea and was awarded the Navy Commendation Medal for leadership and the Southeast Asia Service Medal for Iraq operations in the Red Sea. He earned a Master of Engineering Management degree as he completed his military service.

Bruce Brandes—Bruce Brandes has more than 30 years of experience in executive management and entrepreneurial thought leadership to build growth stage technology-based businesses in the healthcare industry.

Bruce's experiences range as a strategist, operator, entrepreneur, investor, fund-raiser and marketer to advance the transformational promise of digital health. A career-long member of ACHE, Bruce currently serves on the Board of Advisors at several innovative companies and mentors future leaders as Entrepreneur in Residence at the University of Florida's Warrington College of Business.

The healthcare innovations to which Bruce has made meaningful contributions include: developing smart care facilities by introducing ambient intelligence and virtual nursing, scaling consumer and provider adoption of virtual care, creating a platform to reinvent the connection between buyers and sellers of digital health solutions, accelerating the shift toward value-based care, helping pioneer and scale the widespread adoption of remote patient monitoring following the introduction of the iPhone and driving early market acceptance of electronic medical records in the 1990s.

Rachael Britt-McGraw—Tennessee Orthopaedic Alliance— Rachael Britt-McGraw, CHCIO, is Chief Information Officer for Tennessee Orthopaedic Alliance. In this role, Ms. Britt-McGraw serves as an innovative change agent and thought leader for the sprawling surgical practice with 21 locations and more than 150 providers. Believing that IT projects and goals must at all times be aligned with business strategy and articulated to executives in business terms, Ms. Britt-McGraw has more than two decades of proven IT and business success across a wide variety of industries. Ms. Britt-McGraw holds a Bachelor's of Science in Computer Science—Systems Analysis and Design, and a second BS in Management, both earned from the University of North Carolina at Asheville. She currently resides in the Nashville, Tennessee, area where she serves as a speaker on topics of Cyber Security. She is a member of HIMSS, WITT and the Nashville Technology Council, and she holds the CHCIO certification from the College of Health Information Management Executives.

Adam Buckley, MD—UVM Health Network—Adam Buckley, MD, MBA is Chief Information Officer, UVM Health Network. Dr. Buckley was appointed to his current position after serving as the interim Chief Information Officer for the Medical Center from mid-2014 until his appointment to the network role in May 2015. Prior to his role as CIO, Dr. Buckley was the Chief Medical Information Officer at the University of Vermont Medical Center. His focus as CMIO was optimization of the medical center's electronic medical record and stabilization of the team that supports the record. Dr. Buckley came to the medical center and health network with 12 years of experience in academic medicine primarily focused on quality, patient safety and graduate medical education. Dr. Buckley served in a variety of leadership roles both at Beth Israel Medical Center in New York City and Stony Brook University Medical Center in Stony Brook, New York. Dr. Buckley received his medical degree from George Washington University Medical Center in Washington, D.C., and his Master of Business Administration from the Isenberg School of Business at the University of Massachusetts. He completed his residency in obstetrics and gynecology at the McGaw Medical Center of Northwestern University in Chicago, IL. He holds two board certifications, one in OB/GYN and a second in clinical informatics.

José Luis Bueno, MD—José Luis Bueno is Head of the transfusion, apheresis and non-transfusional hemotherapy units at Puerta de Hierro University Hospital in Madrid (Spain). He is a specialist in hematology and holds a MSc in Statistics and Molecular Oncology. Dr. Bueno serves as a member of the Transfusion Safety Commission at the Spanish Health Ministry and as a reference Spanish member in the European Expert Panel on Ebola Convalescent Plasma Use. Also, he is the transfusion advisor at the National Accreditation Office on Quality (ENAC. Industry Ministry of Spain). Dr. Bueno is an active scientific author with several publications in peer-reviewed medical journals and is also a reviewer for *Transfusion* and *Vox Sanguinis*, the two most important journals in transfusion medicine. During the past 10 years he was the Head of the Apheresis, Collection and Blood Processing Areas at the Red Cross Blood Donation Center in Madrid. Dr. Bueno is also an entrepreneur, founding a Biotech company in 2013 called PRoPosit Bio, dedicated to developing cellular therapies in humans and for veterinary uses and IT systems for improving transfusion safety.

Hank Capps, MD—Hank Capps is Executive Vice President and Chief Information and Digital Officer for Wellstar Health System. He is also Co-founder and President of Catalyst by Wellstar, an innovation, consumer research and venture firm. Dr. Capps leads Wellstar's technology, data, innnovation and venture organizations with a particular focus on strategic growth and partnerships across the broader healthcare ecosystem. Dr. Capps is a national thought leader and expert in innovation, healthcare venture, digital design, consumerism and emerging technology. He was recently named the 2022 Georgia CIO of the Year® ORBIE® Award in Healthcare. He is passionate about designing the future of healthcare. Dr. Capps serves on multiple boards and is very active in the community.

Leslie Carroll—Lay Health Advisor at Center for Reducing Health Disparities.

Charles Christian—Indiana Health Information Exchange—Charles Christian is Vice President of Technology and Engagement of the Indiana Health Information Exchange (IHIE), the nation's largest Health Information Exchange. Prior to joining IHIE, Mr. Christian served as Vice President/Chief Information Officer of St. Francis Hospital, a free-standing, acute care, community hospital in West Georgia, a position he held for 2.5 years. Before his role at St. Francis, Mr. Christian served as Chief Information Officer for Good Samaritan Hospital in Vincennes, Indiana, a position he held for almost 24 years. Before joining Good Samaritan Hospital, Mr. Christian worked in healthcare IT for Compucare and Baxter Travenol in both management and implementation roles. Mr. Christian started his career in healthcare as a registered radiologic technologist, serving in various radiology roles for 14 years. Mr. Christian studied natural sciences at the University of Alabama in Birmingham and holds a Bachelor of Science in Business Administration from Lacrosse University.

Daniel Clark—Senior Vice President, AVIA, Center for Care Transformation—As a senior healthcare executive with decades of experience, Daniel has an enduring passion for improving healthcare and crafting an industry that serves us all, uses technology as a strategic catalyst and introduces new approaches to old and new problems.

Jaimie Clark—Jaimie Clark is Co-founder of Catalyst by Wellstar and is currently serving as Head of Innovation. She also serves as Director, Innovation and Venture Strategy for Wellstar Health System. In her roles, she leads the day-to-day operations of a leading health system innovation company charged with designing the future of healthcare and driving growth through emerging business opportunities. She is also active in the Atlanta venture community and serves on the selection committee for Engage, an innovation platform and venture fund comprising a partnership of category-leading corporations. Jaimie recently received the Executive Champion Award from Engage for her breakthrough work in innovation with the organization's startups.

Will Conaway—NTT Data—Will Conaway is Consulting Vice President for NTT Data. He was formerly the Chief Growth Officer for Tech Mahindra Health and Life Sciences. Chief Growth Officer focusing on the organization's business growth, advancing technologies and strategic direction. In his career, he has held the titles Chief Information Officer, Chief of Staff and Chief Operating Officer. Concurrent with his executive roles in healthcare, Will teaches at three colleges at Cornell University: Cornell's ILR School, Cornell's Sloan College of Human Ecology and Cornell's Brooks School of Public Policy, working with masters-level students in the fields of leadership, organizational strategy and healthcare strategy. He has been a featured speaker at graduations and events at Cornell University and Kansas State University. Will was a panelist at the *Roundtable on the State of Healthcare and Technology in the U.S.*, hosted by Harvard Business School and athenahealth. He also emceed and moderated the *CIO & CISO Southern California Summit* in 2020. Will's service on boards and councils includes: AEHIT & AEHIA Boards at CHIME, Forbes Technology Council, Chair of the Kansas State University Psychology Sciences Advisory Council, American Heart Association's ELT Heart & Stroke Ball in Los Angeles, Board of Governors at the Federation of American Hospitals, Manifest MedEx Advisory Committee, AT&T's Healthcare Advisory Council, Verizon's HLS Customer Advisory Board, Fast Company Executive Board and the Los Angeles World Affairs Council. Will was a Becker's Hospital Review's *100 Hospital and Health System CIOs to Know* for 2019. He was a recipient of Constellation's Business Transformation 150 Award in 2020 and 2022, being recognized as Exemplar of Leadership in Driving a Digital Future.

Juan Luis Cruz—Puerta de Hierro University Hospital—Juan Luis Cruz, MSc, was CIO at Puerta de Hierro University Hospital in Madrid (Spain) for more than seven years. Named by the Health Information Management Systems Society (HIMSS) as one of the top 50 Healthcare IT leaders in Europe (2018), he holds a master's degree in telecommunications engineering from the Technical University of Madrid and an executive program from IESE Business School, among others. He is a member of the American College of Healthcare Information Management Executives (CHIME), CHCIO certified and founder and member of the board of the HIMSS Iberian Community. At present he is pursuing a PhD in data analytics and clinical decision support in Oncology. Co-author of several peer-reviewed articles, Mr. Cruz's research interests include the application of data science, machine learning and artificial intelligence in the clinical setting to improve outcomes and the value given to patients. Mr. Cruz has extensive experience in the healthcare sector, working as a healthcare IT consultant and leading many innovative projects related with the digital transformation of healthcare, IT strategy and planning, developing and implementation of EMR and digital health solutions at hospitals.

Alison Darcy, PhD—Alison Darcy is Founder and President of Woebot Health. She is a clinical research psychologist and health tech visionary dedicated to creating smart, scalable and accessible mental healthcare solutions. Her work to explore how digital treatments can help solve human problems began more than 20 years ago when she created one of the first online support groups for people with eating disorders. That work led to PhD level studies in psychology at University College Dublin and post-doctoral training at Stanford School of Medicine and with the American Psychiatric Association. At Stanford, Alison also worked with AI pioneer Andrew Ng to explore the intersection of AI and healthcare, leading his Health Innovation Lab in Computer Science. A frequent speaker on digital therapeutics design and development, barriers to care and engagement, Alison has authored more than 40 publications and been awarded research grants and contracts from the NIH, the Davis Foundation and APA. She is also Adjunct Faculty member in the Psychiatry and Behavioral Sciences Department of Stanford's School of Medicine. Alison holds a PhD, an MLitt and BA in Psychology from University College Dublin.

Cynthia Davis—Methodist Le Bonheur Healthcare—Cynthia Davis is Chief Health Information Officer at Methodist Le Bonheur Healthcare in Memphis, Tennessee. She is a recognized technology strategist with more than 35 years of transformational healthcare experience. Her passion is better as well as failure-free patient care and experience. She has led successful implementation, optimization and customer support efforts for several multi-million dollar fast-track business and clinical transformation projects in single and multi-facility organizations. A registered nurse, Cynthia has been a business and informatics executive in single and multiple facility health systems. She is Fellow in the American College of Healthcare Executives and a regular speaker at industry associations, including ACHE, CHIME and HIMSS.

Kevin Dawson—Extrico Health—Kevin Dawson is Chief Information Officer accountable for innovation, technology strategy, product development, deployment, professional services and technology infrastructure. Before joining Extrico in 2022, he served as CIO of Howard University

Hospital, an academic medical center in Washington, DC, where he developed and operationalized the organization's technology transformation program. Kevin has been Certified Healthcare CIO for more than a decade. He transformed the IT organization of multiple healthcare companies, leading to strategic realignment, improved cybersecurity posture, multi-million dollars of cost savings and the development of award-winning new departments. Kevin is CIO of the year 2022 finalist.

Sakshika Dhingra—Humana—Sakshika is a distinguished technology leader working in the Medicaid line of business who develops clinical application strategy and oversees clinical solutions management. For the last decade, Sakshika has led IT solution delivery, product management and transformational initiatives across multiple payor organizations, such as Blue Cross Blue Shield, Optum and AllState, to name a few. The payor solutions developed under Sakshika's leadership were focused on customer-centricity with the sole objective of delivering value to members as fast as possible.

Laurie Eccleston, MPA, BSN, RN, CPHIMS—Inova Health System—Laurie Eccleston is Manager IT Clinical Ancillary Applications at Inova Health System. Laurie has more than 20 years of IT experience designing, developing, implementing and managing clinical enterprise electronic medical information systems (EMR), as well as 26 years of emergency medicine nursing practice. Laurie has successfully utilized technology and process re-engineering solutions to maximize business efficiencies and has a high level of expertise in gathering, analyzing and defining business and functional requirements. Laurie was a co-presenter at HIMSS 2018 conference "Bricks and Mortar of a Telehealth Initiative" and was the co-inventor of a mobile remote patient monitoring application. Additionally, Laurie was a contributor to the original *Voices of Innovation* and a NursePitch winner in August 2020.

Michelle Evans—Michelle obtained her ADN in 1999 and BSN in 2007, both from Walsh University. She earned her master's as an advanced practice nurse in 2010 from Malone University and is board-certified through the ANCC as a family nurse practitioner. Michelle's clinical background is critical care/rapid response and hospitalist medicine, serving as Hospitalist Nurse Practitioner at Aultman Hospital from 2011–2021. She carried her CCRN Certification through the AACN from 2005–2017 and is Fundamentals of Critical Care Support-Trained through the Society of Critical Care Medicine. Michelle discovered her passion for sepsis care while serving as Intensivist Program Coordinator for Aultman Hospital from 2002–2008. She is currently employed at Summa Health System as Sepsis Program Coordinator, a newly created role and program in January 2021.

Michael Fey—Symantec—Michael Fey joined Symantec as President and Chief Operating Officer following the company's acquisition of Blue Coat, Inc. completed in August 2016. Fey served as President and Chief Operating Officer of Blue Coat starting in 2014. In this role, Fey focused on aligning the company's leading web security, encryption management, cloud offerings and advanced threat protection solutions with customer requirements and was responsible for all aspects of sales, marketing, support, services and product management. Prior to joining Blue Coat, Fey served as Chief Technology Officer of Intel Security and General Manager of corporate products for McAfee, part of Intel Security, where he drove its long-term enterprise strategic vision and innovation in the endpoint, network and security analytics segments. An industry veteran, Fey has held leadership positions with Opsware, Mercury Interactive and Lockheed Martin. Fey graduated magna cum laude with a BS in Engineering Physics from Embry-Riddle Aeronautical University.

Helen Figge—President Elect, NYS HIMSS, Chief Strategy Officer, MedicaSoft—Helen is a healthcare transformer and futurist exceling in organizational mission success. Helen has served in three Fortune companies and non-profit organizations securing exponential authority. She has served and continues to serve on several national committees and boards as well as in several senior advisory roles for executive leadership. Helen's career awards are numerous, including Becker's various women acknowledgements in HIT, a FedHealthIT 100 Winner and several service awards in recognition of performance and contributions to the healthcare IT industry. She publishes, lectures and presents on healthcare technology. She holds academic appointments, has a Baccalaureate in Science, Doctor of Pharmacy, Healthcare Administration MBA, and completed a drug information research fellowship, Boston MA. Helen is a passionate career mentor supporting various causes. She guides start-up companies interested in healthcare technology and serves as Executive in Residence, School of Healthcare Business, Massachusetts College of Pharmacy & Health Sciences University, Boston MA.

Jennifer Fox, MBA—Jennifer Fox is Interim Chief Executive Officer for Mary Crowley Cancer Research. In addition, she is a business consultant and entrepreneur with an emphasis in the life sciences industry. Prior to becoming a consultant, Jennifer was Vice President of Research Operations for Baylor Scott & White Research Institute. She was responsible for strategic execution of innovative research taking place in more than 40 medical specialties across the enterprise. Jennifer played an integral role in furthering the innovation of life saving treatments for patients with severe diseases, including cancer.

Jennifer has taken a personal interest in supporting the cancer community through clinical research. Like most, her family has been impacted by cancer, and she believes in making more clinical trials accessible for patients fighting cancer.

She holds a Bachelor of Science from Stephens College and a Master of Business Administration from the University of Texas at Dallas, and she has more than 20 years of experience in leadership, business strategy and operations management. Jennifer has served on several community boards, is active in the tennis community and lives in Dallas with her husband.

Kyle Frantz—I am an entrepreneurial executive who liberally uses buzzwords, infrequently uses punctuation, and seeks out high integrity, hardworking, bright teams and individuals who do what they do well. With a diverse career history at the intersection of operations and innovation I learned the value of relationships from the front desk to the c-suite. As today's business climate changes rapidly trust and collaboration are more critical than ever. I am committed to innovating how we interact—not just what we produce. I am a lover of the deal who thrives on translating complex technologies to digestible and impactful business solutions.

Jon Friedenberg—As Former CEO of Mary Crowley Cancer Research, a non-profit, phase 1 trial clinic that specializes in the most cutting edge, targeted cancer trials, Jon Friedenberg has a deep appreciation for the role, pace and importance of innovation in healthcare. Previously, Jon served as President and COO of MarinHealth in Greenbrae, California. Prior to that, Jon served as VP of El Camino Hospital in Silicon Valley. There he co-founded the Fogarty Institute for Innovation with legendary inventor and physician Dr. Thomas Fogarty.

Richard Gannotta—UC Irvine Health—Richard Gannotta is Former Chief Executive Officer of UCI Health. He oversees Orange County's only academic medical center and all clinical and patient-serving operations, including the UCI Medical Center in Orange and ambulatory sites across the county. Prior to joining UCI, Gannotta was Senior Vice President of hospitals at New York's NYC Health + Hospitals, the nation's largest public healthcare system. Before that, he was President of Chicago's Northwestern Memorial Hospital, where he led hospital operations and worked closely with the leadership of Northwestern University's Feinberg School of Medicine. He has also served as President at Duke Raleigh Hospital, part of the Duke University Health System and North Carolina-based WakeMed Health and Hospitals. Gannotta's areas of research interest include the economic impact of alternative models of care, patient safety, high-reliability systems and healthcare disrupters. His academic affiliations include teaching at the undergraduate and graduate levels, and he is currently Visiting Scholar at NYU Wagner School.

John Goodgame—InterSystems Corporation—John Goodgame is Senior Sales Engineer at InterSystems Corporation, mentoring and supporting some of the largest health systems in the world with interoperability, design and development. John has more than 30 years of experience with healthcare technology working in clinical, hospital and vendor environments. He has held positions in healthcare that include developer, manager, consultant and director and continues to enjoy solving problems presented in the modern healthcare industry.

Shawntea Gordon—MBA FACMPE CMOM—H4 Technology, LLC— Shawntea (Taya) Gordon is Chief Revenue Cycle Officer at H4 Technology, author of the MGMA Publication "Revenue Cycle Management: Don't Get Lost in the Financial Maze," host of the podcast "RevDive" and general revenue cycle management enthusiast. Taya received her Executive Healthcare MBA from Creighton University and has held senior level executive management positions in private practices, collaborative institutes and national care coordination organizations. Taya now educates and consults on all areas of healthcare operations and the opportunity for improvements through greater use of technology. Taya is a proud member of MGMA's Government Affairs Council where she displays her passion for physician advocacy, community outreach and healthcare improvement processes. Taya contributed to the first edition of *Voices of Innovation*, serves as Adjunct Instructor for the Practice Management Institute and spends her spare time with Habitat for Humanity Omaha.

Alex Goryachev—Cisco—Alex Goryachev is an entrepreneurial go-getter who loves blazing new trails for innovation. Over the past 20 years, he's turned disruptive concepts into emerging business models. His passion is to create a strategy and then drive it home to "get things done." As Cisco's Managing Director of Innovation Strategy and Programs, he has plenty of opportunities. First, Alex spearheads programs inspiring all employees to share their big ideas, many of which the company helps to codevelop. He carries the torch for co-innovation across the company's partner ecosystem via Cisco's Innovation Centers and Grand Challenges, and he works hand-in-hand across functions to spot and monetize emerging technology and business transitions. Prior to joining Cisco in 2004, Alex consulted at Napster, Liquid Audio, IBM Global Services and Pfizer Pharmaceuticals. Alex was the Emerging Stars Gold winner of Brandon Hall Group's 2016 Human Capital Management Excellence Awards, and his organization also won Golds for best Innovative Talent Management and Employee Engagement programs. A sought-after keynoter and media authority on innovation, Alex has a passion for sharing knowledge, mentoring and guiding innovation programs.

Sarah Hatchett—Sarah Hatchett is a results-driven leader with more than 18 years of experience of healthcare IT leadership, primarily focused in the PMO and EHR practice domains. She has been at Cleveland Clinic for the last five years and is currently serving as Associate Chief Information Officer leading ITD's Division Administration and Growth functions. She started her career at Epic where she participated in more than 30 implementations, including some of the fastest, innovative and most challenging projects that helped shape Epic's current implementation methodology. As a consultant, she worked with Sutter Health in Sacramento and New York City Health + Hospitals to develop complex rollout strategies, create robust EHR optimization and upgrade programs and tailor supporting PMO methodologies. She has led implementations including a diverse portfolio clinical and revenue cycle applications, hardware, infrastructure and interfaces. Her current portfolio includes oversight of more than $500M of projects, health system integrations, new hospital constructions, routine capital projects and IT operations. Sarah graduated with a BA in English and Creative Writing from UW-Madison and achieved her MBA from Baldwin Wallace University. She is Project Management Professional (PMP) certified and once won the best smile contest at the Wisconsin State Fair.

Steve Hess—Steve has been working in the health care IT field for more than 31 years with 19 years as Chief Information Officer. Steve was named CIO of UCHealth (University of Colorado Health) in 2012. UCHealth is a Colorado-based healthcare system that includes 12 hospitals, including the Rocky Mountain region's only academic medical center, University of Colorado Hospital and hundreds of clinics throughout Colorado, Wyoming and Western Nebraska. Steve and his team have focused on creating a consolidated, enterprise IT platform across UCHealth, laying the foundation for innovation and transformational care. The UCHealth system has achieved HIMSS Stage 7 and Most Wired Level 10 for both acute and ambulatory. In addition to the integrated IT platform background, Steve has extensive experience with digital patient experience capabilities, health information exchange, advanced clinical and research analytics, mobile technology, personalized medicine and virtual health.

Dan Howard, MBA, PMP, CHCIO, CDH-E—Dan Howard is Vice President and Chief Information Officer for San Ysidro Health. He leads various teams within the organization and is responsible for all technologies, application development, clinical and business applications and digital transformation efforts.

Before joining San Ysidro Health, Mr. Howard ran his own consultancy, served as Chief Information Officer at the University of South Alabama Heath System and Site CIO for Loma Linda University Health, and held various technology leadership positions at Fidelity Information Systems, Metavante, Vicor, and Kaiser Permanente. Prior to his technology career, Dan worked on the frontlines of healthcare for nine years, specializing in pediatric and neonatal intensive care nursing.

Dan holds undergraduate degrees in nursing and biology and has an MBA with an information management and finance emphasis. He maintains active certifications as a project management professional is Certified Healthcare CIO and Certified Digital Health Executive through the College of Health Information Management Executives.

Amy Hushen—Johns Hopkins School of Medicine—Amy Hushen holds a BFA in graphic design from the Maryland Institute College of Art. She leads graphic design work for the Technology Innovation Center (TIC), including branding identity design, print/marketing collateral and user interface design. At the TIC, Amy focuses on the functionality and overall visual elements that improve the customer's ability to navigate within a digital product.

Gary Johnson, PharmD, MHA—University of Kentucky Healthcare—Gary Johnson, is Chief Innovation Officer at the University of

Kentucky Healthcare. Dr. Johnson's research interests include the assessment of ambulatory pharmacy services, including quality, access and profitability. His recent work has focused on the expansion of ambulatory pharmacy services. Dr. Johnson received his PharmD and MHA degrees from Mercer University. He completed two postgraduate residency programs, including a management focused residency at the University of Cincinnati. Dr. Johnson served as Assistant Director of Pharmacy at the University of Kansas Medical Center and Director of Pharmacy at the University of Virginia Health System. He joined the University of Kentucky in 2011 as Chief Pharmacy Officer and transitioned to the role of Chief Innovation Officer in 2017.

Tarun Kapoor, MD, MBA–Virtua Health—Tarun is Senior Vice President and Chief Digital Transformation Officer at Virtua Health. In this role, he oversees Virtua's Digital Transformation Office and orchestrates Virtua's enterprise-wide master plan in support of an intuitive care journey for all consumers. Previously, Dr. Kapoor was President of VirtuaPhysicianPartners™ and Senior Vice President and Chief Medical Officer for Virtua Medical Group (VMG), a clinician multi-specialty medical practice.

Inderpal 'Inder' Kohli—Inder Kohli serves as Vice President of Information Technology and Chief Information Officer at Englewood Health in New Jersey, where he oversees all technology and digital solutions and strategies for the health system, including its acute care hospital and a network of more than 140 locations, including physician offices, hospital services and urgent care centers, all connected through one electronic medical record system. Kohli has more than 20 years of experience as a leader in the information and healthcare technology field, with extensive experience planning and implementing enterprise information systems to support centralized clinical and business operations. Prior to Englewood, Kohli served as Assistant Vice President at the Hospital for Special Surgery in New York City, where he led the implementation of several enterprise information systems, including Epic, to support clinical and business operations. Kohli also teaches at the master's level. He oversees a curriculum on healthcare informatics at Weill Cornell Medical College in New York; it is here that he transfers his knowledge to the next generation of IT professionals.

Krishna Kurapati–QliqSOFT—Krishna is Founder and CEO of QliqSOFT. He has more than two decades of technology entrepreneurship experience. Krishna started QliqSOFT with the strong desire to solve clinical collaboration and workflow challenges using artificial intelligence (AI)-powered digital technologies across the U.S. healthcare system. He is actively involved in early-stage financing of startups in both the U.S. and India.

Gigi La Course, MT ASCP—Inova Health System—Gigi La Course is Team Lead of the Inova Integration team at Inova Health System. Gigi has 28 years of IT experience in various types of clinical and non-clinical applications as well as 10 years of medical technology experience. She has designed, implemented and supported laboratory, electronic medical records, time and attendance, nurse scheduling and integration engine applications. During the last 12 years, Gigi has been integral in the design and development of what is now InterSystems Iris for Health (integration engine) as well as Epic's Bridges module at Inova Health System. She has coded, tested and implemented and currently supports hundreds of clinical integration projects that have positively affected patient care and streamlined the workflow of clinical practitioners.

John Lee, MD—John Lee is Emergency Physician and Clinical Informaticist. Over the past decade, he has held chief medical information officer positions at two multihospital systems. His vision is a healthcare system that is driven completely by transparent data, information and knowledge, delivered efficiently. Dr. Lee is a member of several professional health IT organizations, including AMIA, AMDIS, CHIME and HIMSS. He has served on several committees within these organizations and within the Meditech and Epic vendor communities. In 2019, he was honored with HIMSS's prestigious Physician Executive of the Year award.

Dr. Christopher Longhurst—UC San Diego School of Medicine and UC San Diego Health—Chris now serves as Chief Medical Officer but has all of technology reporting to him. Chris wrote this while Chief Information Officer. Dr. Longhurst is responsible for all operations and strategic planning for information and communications technology across the multiple hospitals, clinics and professional schools that encompass UC San Diego Health. Dr. Longhurst is also Clinical Professor of Biomedical Informatics and Pediatrics at UC San Diego School of Medicine and continues to see patients. As a result of his efforts to leverage technology to improve patient experience in the UCSD Jacobs Medical Center, he was voted the 2017 Top Tech Exec in the education category for San Diego. He previously served as Chief Medical Information Officer for Stanford Children's Health and Clinical Professor at the Stanford University School of Medicine, where he helped lead the organization through the implementation of a comprehensive electronic medical record (EMR) for more than a decade. This work culminated in HIMSS Stage 7 awards for both Lucile Packard Children's Hospital and 167 network practices in Stanford Children's Health.

Emily Marx—Johns Hopkins School of Medicine—Emily Marx serves as Communications Coordinator for the Technology Innovation Center and holds a BS in Psychology from Towson University. Emily organizes TIC outreach efforts such as e-newsletters, copy writing, events planning and social media promotion. As a member of the TIC design team, she assists in the planning of user research and product marketing. Emily also has a hand in the coordination and management of TIC's leadership development programs focused around entrepreneurship, analytics and precision medicine.

Jasmine McNeil—Johns Hopkins School of Medicine—Jasmine McNeil leads marketing and design work at the Technology Innovation Center, Johns Hopkins Medicine. She holds an MA/MBA in Design Leadership from MICA and the Johns Hopkins Carey Business School. At the TIC, Jasmine runs design thinking-facilitated sessions to gather user requirements for the applications the TIC builds. She also organizes the TIC's accelerator program (Hexcite) for early-stage medical software startup ideas. Jasmine brings Hexcite participants through a series of technical and business design activities to create project teams and startups.

Joey Meneses—Joey Meneses is Chief Technology Officer at Akron Children's Hospital. He is a digital technology executive, strategist and visionary thought leader—a leader who combines business acumen with digital expertise to help organizations drive digital transformation via an enterprise-wise digital vision and strategy. Joey is a forward-thinking technology executive with years of success transforming large businesses with the adoption of progressive technologies. He offers 20+ years of expertise in driving enterprise capabilities by leading IT vision, strategy and large-scale project initiatives and excels at transforming enterprise applications, infrastructure, service management, data engineering and analytical systems. Throughout his highly decorated career, he was U.S. Air Force Officer and served various roles, including Senior Staff Advisor to the Department of Defense Military Health System on strategic matters related to IT policy, procedures, procurement, solutions and IT operations worldwide and evaluation of various technologies. He led and managed delivery of multiple mission critical systems, operations and systems integration in various state government, U.S. Department of Defense (DoD), commercial, and health-care organizations in the United States, Turkey, Germany, the Philippines and Japan.

Kristin Myers—Mount Sinai Health System—Kristin Myers is Senior Vice President in Information Technology who oversees application strategy, the Clinical Application and Interoperability portfolio, and the IT Program Management Office for the Mount Sinai Health System. For the last 14 years, Kristin has led the transformation of the Epic clinical and revenue cycle implementations across the health system. Under Kristin's leadership, Mount Sinai was awarded the prestigious HIMSS 2012 Enterprise Davies Award of Excellence for its electronic record implementation. The Mount Sinai Hospital and Mount Sinai Queens have achieved HIMSS Stage 6 certification, and will be seeking Stage 7 certification this year. Kristin, an active member of the Corporate Diversity Council, also leads the Women in IT team, which meets regularly to educate and advocate on gender parity. Kristin previously worked in the healthcare technology vendor and consulting space before joining Mount Sinai Health System.

Paul Nagy, PhD, FSIIM—Johns Hopkins School of Medicine—Paul Nagy is Associate Professor of Radiology at the Johns Hopkins University School of Medicine. Dr. Nagy serves as Deputy Director of the Johns Hopkins Medicine Technology Innovation Center (TIC) with the goal of partner-ing with clinical inventors to create novel clinical IT solutions. This team of designers, developers and data scientists work with inventors to build, deploy and evaluate digital health solutions within the Johns Hopkins Medical System. At Johns Hopkins, Dr. Nagy serves as Program Director for multidis-ciplinary leadership development programs in precision medicine, clinical informatics, data science and creating commercial ventures; 360 faculty and staff have taken these programs since 2012. Dr. Nagy is the author of more than 130 papers in the fields of informatics and implementation science.

Aaron Neinstein, MD—Aaron Neinstein is Vice President Digital Health at the University of California San Francisco (UCSF) Health, Senior Director at the UCSF Center for Digital Health Innovation, and Associate Professor of Medicine at UCSF. He has led development and implementation of a wide range of health technology, focusing on improving patient access and expe-rience, as well as new virtual care models. Dr. Neinstein was an inaugural inductee as Fellow of the American Medical Informatics Association (FAMIA) and was appointed to serve on the United States Health Information Technology Advisory Committee (HITAC). He is also a practicing endocrinologist focused on diabetes care. He received his Doctor of Medicine from the Keck School of Medicine of the University of Southern California.

Pete O'Neill—Cleveland Clinic—Pete O'Neill was Executive Director of Cleveland Clinic Innovations, where he managed a group of more than 50 professionals responsible for commercializing Cleveland Clinic's IP portfolio. He had been with Cleveland Clinic Innovations for 13 years, in roles including Director of Commercialization and Senior Officer for commercialization of orthopedic technologies. Mr. O'Neill is passionate about promoting entrepreneurial efforts within Cleveland Clinic's commercialization ecosystem. Mr. O'Neill was educated as an aeronautical engineer at MIT and previously held management positions in aerospace manufacturing and technology commercialization.

Anna Pannier, FACHE, CHCIO—Anna Pannier is Vice President of Information Technology with Centauri Health Solutions where she oversees integration, engineering, service excellence and automation. Centauri is a tech-enabled services and data liquidity company serving health plans and providers throughout the United States. For the past 20+ years, Anna has helped healthcare provider and technology organizations grow, innovate and adopt change. Anna teaches leadership skills to physicians through the Tennessee Chapters of American College of Healthcare Executives and previously taught healthcare data analytics and analysis at Lipscomb University. Anna actively volunteers with ACHE, Women in Technology Tennessee and Tennessee prison ministry and is an active member of CHIME and Nashville Technology Council.

Mitchell Parker, CISSP—IU Health—Mitchell Parker is Executive Director, Information Security and Compliance, at IU Health in Indianapolis, Indiana. Mitch is currently working on redeveloping the information security program at IU Health and regularly works with multiple non-technology stakeholders to improve it. He also speaks regularly at multiple conferences and workshops, including HIMSS, IEEE TechIgnite and Internet of Medical Things. Mitch has a bachelor's degree in computer science from Bloomsburg University, an MS in information technology leadership from LaSalle University and his MBA from Temple University.

Brittany Partridge—Brittany Partridge leads the Virtual Care Technical team at UC San Diego Health, focusing on standing up architectural governance to align infrastructure to support growth of Telehealth, RPM and Virtual Care programs. She has served as Clinical Informaticist for the past 10 years, at Ascension and UC San Diego Health, managing the CI portfolio, bringing hospitals up from paper and growing virtual care programs. She began her career in Fire/EMS and found her love to bringing technology to support clinical workflow at the California Emergency Medical Services Authority. Brittany has been a long-time member of AMIA and has served on the Public Policy Committee and recently completed the Women in AMIA Leadership Program. Brittany is very involved in volunteer opportunities with both Remote Area Medical and Team Rubicon, leveraging her technical skills to support disaster response/mitigation and provide healthcare to the underserved. When she is not working, Brittany enjoys programming and the outdoors, where she is happiest on, in, or by the water.

Michael A. Pfeffer, MD, FACP—Michael A. Pfeffer serves as Chief Information Officer and Associate Dean for Stanford Health Care and Stanford University School of Medicine. Michael oversees Technology and Digital Solutions (TDS), responsible for providing world-class technology solutions to Stanford Health Care and School of Medicine, enabling new opportunities for groundbreaking research, teaching and compassionate care across two hospitals and more than 150 clinics. Michael also serves as Clinical Professor in the Department of Medicine and Division of Hospital Medicine with a joint appointment in the center for Biomedical Research (BMIR). Prior to joining Stanford Medicine, Michael served as Assistant Vice Chancellor and Chief Information Officer for UCLA Health Sciences. During his tenure, Michael was the lead physician for the largest single-day electronic health record go-live encompassing more than 26,000 users and subsequently the Chief Medical Informatics Officer before becoming CIO.

Marc Probst—Intermountain Healthcare—Marc Probst is Former Chief Information Officer and Vice President at Intermountain Healthcare, a not-for-profit, integrated delivery network of 22 hospitals, 185 clinics, a health plans division and other healthcare-related services based in Salt Lake City, Utah. Probst is nationally recognized as a CIO and served on the Federal Healthcare Information Technology Policy Committee, which assisted in developing HIT Policy for the U.S. Government. For more than 30 years, Probst has been involved in healthcare and technology. Prior to

Intermountain, Probst was a partner with two large professional service organizations, Deloitte Consulting and Ernst & Young, serving healthcare provider and payer organizations. Probst has significant interest in the use of information technology to increase patient care quality and to lower the costs of care. He is experienced in information technology planning, design, development, deployment and operation, as well as policy development for HIT-related issues. Marc received a degree in finance from the University of Utah and later a Masters of Business Management from George Washington University. Marc and his wife now service as full-time missionaries in South America.

Dwight Raum—Johns Hopkins School of Medicine—Dwight Raum is Vice President and Chief Technology Officer of Johns Hopkins Health System and Johns Hopkins University. Raum has been with Johns Hopkins more than 17 years and is currently serving as a leader in IT infrastructure operations, product innovation, university information systems and precision medicine. His passion lies in challenging the status quo, mobilizing teams to harness technology and championing change. In 2014, Raum co-founded the Technology Innovation Center, which cultivates innovative faculty and teams them with technical experts to solve problems. He now serves as its Executive Director. Raum also leads technical platform implementation of the Johns Hopkins precision medicine initiative called InHealth, which combines research, data science, technology and clinical disciplines into an integrated program that is transforming the standard of care into precision medicine. Raum holds a bachelor's degree in management science from Virginia Tech.

Kathy Ray—Kathy Ray has been Caregiver for the Cleveland Clinic for 19 years. She has served in multiple roles with the most current position of Patient Service Specialist. During her tenure, Kathy served as a member of the Cleveland Clinic Main Campus Our Voice Healthcare Partner Council and Co-Chair for the regional Medina Hospital Our Voice Healthcare Partner Council. Kathy co led the ITD Initiative to create an IT Our Voice Healthcare Partner Council. Kathy was the recipient of the 2011 Cleveland Clinic Individual Caregiver Award, 2018 member of the Business Technology Leadership Academy, Member of IT Caregiver Council and Toastmasters. Kathy was also the recipient of the Cleveland Clinic 2021 Innovations Award for MyRoad—My Recordable On Demand Audio Discharge.

Stephanie Reel—Johns Hopkins School of Medicine—Stephanie Reel is retired Chief Information Officer for all divisions of the Johns

Hopkins University and Health System. She was appointed Vice Provost for Information Technology and CIO for Johns Hopkins University in January 1999. She is also Senior Vice President for Information Services for the Johns Hopkins Health System and Senior Vice President and CIO for Johns Hopkins Medicine, positions she has held since 1994. As CIO, Ms. Reel leads the implementation of the strategic plan for information services, networking, telecommunications as well as clinical, research and instructional technologies across the enterprise. Ms. Reel graduated from the University of Maryland with a degree in information systems management and holds an MBA from Loyola College in Maryland.

Suzanne Richardson, MSN, RN—Medical University of South Carolina—Suzanne Richardson is Program Manager for the Structural Heart and Valve Center and Cardiac Surgery program at the Medical University of South Carolina. Her professional interests include clinical care logistics and using technology to improve quality of care and the patient experience. Suzanne has more than 16 years of diverse nursing experience in both the private and academic sectors. Suzanne has a high level of expertise in sculpting clinical infrastructure to support concierge service for highly complex patient populations. Additionally, Suzanne is the inventor of a mobile remote patient monitoring application that was highlighted in her co-presentation of "Bricks and Mortar of a Telehealth Initiative" at the HIMSS 2018 conference. She has also presented her patient-facing mobile application and nurse facilitated program as poster presentations at several other local and regional conferences.

Leah Rosengaus, MS—Leah Rosengaus is Director of Digital Health Care Integration at Stanford Health Care. In her role, she is accountable for the design, implementation, scaling and evaluation of strategic digital health services for the health system, including provider-to-provider and provider-to-patient solutions. Prior to joining Stanford Health Care, Rosengaus directed the telehealth program at UW Medicine where she developed virtual care offerings such as asynchronous eConsults, TeleStroke, TeleOB, TelePsychiatry and an on-demand virtual clinic. She has been active in policy work over the last 10 years, advocating for telehealth coverage, reimbursement and access at the state and federal level.

Bill Russell—Health Lyrics—Bill Russell is Founder of Health Lyrics, a management consulting firm that moves health systems to the cloud. Bill has served on executive teams in healthcare, higher education and consulting practices over the past 30 years. While Chief Information Officer for St. Joseph Health, a 16-hospital $5 billion system, he rapidly accelerated the diffusion of new IT strategies and methods and dramatically improved operational effectiveness while also creating new revenue streams by investing in successful startups such as Hart and Clearsense. Bill holds a bachelor's degree in economics from Moravian College in Bethlehem, Pennsylvania, and has completed executive education courses and healthcare IT leadership training at the Harvard School of Public Health in Cambridge, Massachusetts.

Carla Samà—KPMG Advisory Italy—Carla Samà is Manager of KPMG Italy and works for the Healthcare & Life Sciences business unit. She is part of a team of more than 130 healthcare consultants who make themselves available with the aim of boosting consistent and sustainable innovation in the National Healthcare System. For the last eight years, Carla has taken part of healthcare projects ranging from drawing up business plans for healthcare providers, process reengineering and digitalization, management support to Italian Regions in recovery plans in cost containing and planning investments, healthcare personnel sizing and quality of care projects.

Rosie Sanchez—Assistant Vice President | Information Technology Texas Tech University Health Sciences Center

Drew Schiller—Validic—Drew Schiller is CEO and Co-founder of Validic, Inc. Drew's mission is to improve the quality of human life by building technology that makes personal data actionable. In service of this mission, and as a patented technologist, Drew regularly speaks and writes on how technology can humanize the healthcare experience and create an invisible, data-driven system. Drew serves as a member of the Consumer Technology Association's (CTA) Board of Industry Leaders, as well as Vice Chair for CTA's Health and Fitness Technology Division Board. Through his work with CTA, he contributes to moving the industry forward by developing data standards that advance performance benchmarks and consumer acceptance. Additionally, Drew serves as an advisor for the Clinical Trials Transformation Initiative, as a board member for the Council for Entrepreneurial Development, and as a member of the eHealth Initiatives' Board of Directors. Most recently, Drew was recognized by MedTech Boston as a 40 Under 40 Healthcare Innovator for his work to progress the use of data and technology in healthcare.

Jonathan Scholl—Leidos—Former CEO of Leidos Healthcare, Jon and his wife now serve as full-time missionaries.

Roberta Schwartz, PhD, MHS—Roberta Schwartz is Executive Vice President and Chief Innovation Officer of Houston Methodist Hospital, one of the Texas Medical Center's founding institutions. She is responsible for overseeing all operations at the 946-bed hospital, which has been named by *U.S. News & World Report* as the No. 1 hospital in Texas for 11 straight years and has also been named to the publication's prestigious "Honor Roll" of America's best 20 hospitals six times. In her role as CIO, Roberta is responsible for advancing and expanding Houston Methodist's digital innovation platforms, including telemedicine, artificial intelligence and big data.

Prior to joining Houston Methodist, Roberta worked as Director of Business Development for Mount Sinai School of Medicine in New York and as Consultant and Project Manager for several academic medical centers for APM/Computer Sciences Corporation and for CMS (HCFA).

Roberta earned a master's in health science from Johns Hopkins University and an honors undergraduate degree from Barnard College at Columbia University. She has a PhD from the University of Texas School of Public Health. Roberta has been recognized nationally for her professional and non-profit work. She was most recently recognized by Modern Healthcare's Top 25 Women Leaders as one of its 10 Women to Watch. Roberta is also involved in many non-profit organizations such as Lifegift OPO, Robert M Beren Academy and the UOS Synagogue, Young Survival Coalition and many breast cancer organizations.

Amy Sheon, PhD, MPH—Case Western Reserve University School of Medicine—Amy Sheon is Executive Director of the Urban Health Initiative at Case Western Reserve University School of Medicine. Dr. Sheon has worked on a variety of emerging public health issues, from HIV prevention in the epidemic's earliest years to ethical issues in genetics and childhood obesity. Her current focus is on ensuring that advances in health information technology can reduce rather than accentuate health disparities. Thus, she is developing and testing a model of universally screening patients for their digital skills and connectivity, referring to local partners to address gaps and then training patients to use portals, remote monitors, telehealth and apps. Sheon earned her PhD at Johns Hopkins Bloomberg School of

Public Health and her MPH at the University of Michigan. Previously, she worked at the National Institutes of Health, the University of Michigan and Altarum Institute.

Christopher "Topher" Sharp, MD—Topher Sharp serves as Chief Medical Information Officer at Stanford Health Care, where he is a physician leader in the application of innovative clinical technologies for the delivery of care. Dr. Sharp is accountable for integrating and transforming health technologies to promote the best clinical care experience and clinical outcomes for patients at Stanford HealthCare. He has responsibility to ensure that clinical information systems work well for clinicians—and that clinicians are proficient and effective in their use of these systems in the care of their patients. He drives the clinical strategy, design, enhancement, usability, adoption and workflow integration of clinical information systems and digital health for Stanford Health Care. As Clinical Professor in the Department of Medicine, Dr. Sharp maintains an active primary care and inpatient practice. He has published multiple academic articles on the impacts of clinical informatics in safety, outcomes and clinician wellness. Dr. Sharp serves as a mentor and educator for Stanford Clinical Informatics Fellows, junior faculty and other trainees.

Stefanie Shimko-Lin, RN—Cleveland Clinic—Stefanie Shimko-Lin serves as Clinical Analyst under the Digital Domain Patient Journey and Clinical Delivery Team at the Cleveland Clinic. The team provides innovative solutions to clinical issues using an agile design process. Stefanie's recent focus, the homegrown mobile application Iris Mobile and Epic Mobility, leverages mobile technology to provide end users the ability to review patient data and document on the go. End of life is an area of special interest to Stefanie. The Clinical Delivery Team utilizes existing technology to create tools to visually prioritize patient care and document in real time, improving patient safety and accuracy in electronic documentation. Stefanie has served as a nurse for more than 10 years with clinical focus areas of child birth, gynecology, forensics, rehabilitation and end of life.

Angela Skrzynski, MD–Virtua Health—Angela is board certified in family medicine and sleep medicine, having completed her training at St. Joseph's Hospital in Philadelphia and the University of Michigan. She serves as Lead Physician for Telehealth at Virtua Health. She has been with the team since its inception and is responsible for clinician oversight and training, maintaining quality and safety and innovating with the digital transformation team, as well as direct patient care via telemedicine. She lives in southern New Jersey with her husband and two little girls.

Gregory Skulmoski—Bond University—Greg Skulmoski is Assistant Professor at Bond University and leads the Master of Project Innovation program. He has 15 years of project experience and 10 years teaching project management in North America, the Middle East and Australia. He has published more than 40 peer-reviewed papers in leading journals. Greg's research interest is in innovation. His PhD is from the University of Calgary. Greg also served as Executive Consultant for Accenture in Australia and the United Kingdom.

Josh Sol—Josh Sol currently serves as Managing Director at FTI Consulting leading the digital health solutions team focusing on digital modernization and emerging technology. With 15 years of experience in healthcare information technology and digital transformation, he is passionate about empowering health strategies with technology and is focused on refining the digital experience for patients and clinicians. Josh has worked to establish a reputation as an innovator and strategist, team builder and a forward-thinking leader in the HIT industry. He has a proven track record of aligning innovation and IT strategy with operational needs, ensuring transformational impacts to care delivery. Before joining FTI Consulting, Josh was Administrative Director at Houston Methodist, advising, implementing and optimizing integrated electronic medical records systems, ultimately achieving HIMSS Stage 7 certification, Most Wired 9 in the ambulatory/outpatient space. He specialized in interoperability, population health and patient experience technologies and innovative product development opportunities and was a key leader in the Epic Implementation, Community Connect program, Center for Innovation as well as the Technology Hub.

Amy Szabo—Cleveland Clinic—Amy Paige Szabo is Continuous Improvement Consultant on the Heart, Vascular, & Thoracic Institute's Advisory Team. Amy Szabo specializes in a lean continuous improvement approach, focusing on best practices, efficiency, cost containment, systematic processes and patient safety. She has 26 years of experience in healthcare and operations and has demonstrated the ability to deal effectively with people in the supervision of staff and team members. She successfully oversaw and implemented changes in approach, methods, products

---

and processes to enhance competitiveness and improve organizational efficiencies. Amy is a firefighter, emergency department caregiver and nurse leader.

Peter Tippett—CEO and Founder of careMESH, Dr. Tippett was previously the Chief Medical Officer of Verizon and VP of the Innovation Incubator. Among other technology start-ups, Tippett created the first commercial anti-virus product, which became Norton Antivirus; founded TruSecure and CyberTrust and was Chairman of MD-IT. He was a member of PITAC (President's Information Technology Advisory Committee), which drove modern Health IT transformation policy, served 3 years with the Clinton Health Matters initiative and recently served on the Whitehouse/NIH Precision Medicine Initiative.

Dr. Tippett is a recognized expert and thought leader in enterprise information security, risk and compliance as well as in Health Information Technology and health technology transformation. He is a serial entrepreneur, a senior executive and frequent speaker who has helped numerous boards-of-directors and Fortune-500 leaders to understand and build successful approaches to security, risk, compliance, health technology transformation and accelerating innovation.

Steve Trilling—Symantec—Steve Trilling is Senior Vice President and General Manager of Security Analytics and Research at Symantec, the global leader in cyber security. Symantec helps companies, governments and individuals secure their most important data wherever it lives, from mobile devices and the connected home, to desktops and laptops, to enterprise data centers and the cloud. Steve's division delivers Symantec's industry-leading threat protection technologies, advanced security analytics, investigations into new targeted cyberattacks and breakthrough innovations in artificial intelligence and machine learning, as well as a variety of shared services including product localization and product security. Trilling holds a BS in Computer Science and Mathematics from Yale University and an MS in Computer Science from the Massachusetts Institute of Technology.

Meera Udayakumar, MD, SFHM—Dr. Udayakumar serves as Medical Director for Advanced Care at Home and Executive Medical Director of Quality and Innovation for Triangle East. She is Hospitalist at UNC Rex, where she also serves as Chair of Medicine for Rex Physicians. Dr. Udayakumar has been named a senior fellow in hospital medicine and has completed advanced leadership training through the Center for Creative Leadership, as well as the North Carolina Medical Society. Her current professional interests include artificial intelligence and data science in healthcare. Her Advanced Care at Home program was recently awarded the 2022 Highsmith Award for Innovation, presented from the NC Healthcare Association.

Marissa Ventura, MS—UC San Diego Health—Marissa Ventura is Communication Lead for Information Services at UC San Diego Health. She is a strategic communicator with experience in the healthcare and pharmaceutical industries and has worked as a medical journalist and editor. She has an MS from The Graduate School of Journalism at Columbia University.

Sergio Villegas—Murata—Also with Murata's Corporate Technology & Innovation group, Sergio Villegas is Senior Manager, Startup Incubation & Investment. He brings nearly 20 years of experience across multiple functions (engineering, finance, marketing, incubation, strategy and investments) and industries (manufacturing, healthtech, transportation, and financial services and SaaS). In his current position, he interfaces with more than a dozen R&D and product business units to better understand customer and technology needs and gaps and bridges those gaps with external startups, corporations and universities. Sergio graduated from San Diego State University with a BS in Mechanical Engineering and earned an MBA from the University of California Berkeley.

Donna Walker—Revenue Integrity Audit—Donna Walker is Nurse Coder Analyst and Consultant with expertise in revenue cycle and program integrity. She has been recognized for detailed findings of fact in medical records and claims with Medicare, Medicaid and commercial payors.

John Walker—Long time healthcare transformation leader. Seasoned nurse coder analyst and consultant; with expertise in revenue cycle and program integrity. Recognized for detailed findings of fact in medical records and claims; with Medicare, Medicaid and commercial payors.

Mark Waugh is Senior Manager, Healthcare in the Corporate Technology & Innovation group at Murata Electronics North America, Inc. With more than 25 years of diverse global experience, his previous roles include strategic marketing, product management and process engineering in demanding, highly competitive consumer technology environments.

His leadership skills, coupled with strong communication and relationship management expertise, enable engagement among all levels of management, programs and processes. As a result, the initiatives that Mark spearheaded have resulted in expanded collaboration opportunities, maximized sales channels and increased market penetration. He holds a BS in Materials Science and Engineering from Penn State University.

Danielle Wilson–Virtua Health—Danielle is Assistant Vice President of Digital Transformation at Virtua Health with more than 20 years of information technology and healthcare experience. In her current role, she focuses on the enterprise expansion of how digital technologies influence and support healthcare imperatives. This includes the strategy and operations of a digital transformation office that supports a portfolio of technology and analytical capabilities at Virtua. These efforts lead to a culture of business insights, opportunity realizations, operational changes and data-driven decision-making.

Charles Zonfa, MD—Dr. Charles Zonfa is SVP/Chief Quality Officer serving as Senior Clinical Executive of Summa Health, where he leads and reports all clinical quality and patient engagement performance across the health system, including patient safety and clinical resource management for all Summa entities, including all hospitals, clinical service lines, medical group, homecare/hospice, ambulatory (including surgery) sites, ACO and the health plan. Dr. Zonfa has focused on driving Summa's population health strategy by integrating clinical services and analytics across the organization. This strategy has been a key component in Summa's health plan, SummaCare, achieving a CMS 5 Star Rating. Dr. Zonfa is a board-certified obstetrician/gynecologist with more than 20 years of experience in the healthcare industry as both a clinician and healthcare executive.

Book Introduction

Edward W. Marx

I have preserved the original Introduction here as it was written when *Voices*, first edition was published in 2019. I only want to add one paragraph to *Voices*, second edition.

Shortly after publishing *Voices*, I was diagnosed with late-stage prostate cancer. The journey is well chronicled in other spaces should you have an interest, but here is the bottom line. My life was saved, and my recovery made complete, directly through innovation. Instead of relying on traditional prostate cancer blood tests that are 50% accurate, I was administered an artificial intelligence (AI) powered lab test that has 90% accuracy. There was no doubt as to the validity of my diagnosis and the urgency to get treated quickly. In recovery, we leveraged digital tools to ensure I and my family had an optimal experience. Thanks to excellent clinicians and innovative tools, I remain cancer free three years later. This is why I am more than thrilled to put in the effort to update *Voices* with the hope that readers and their organizations will be both inspired and educated with best practices to ensure a culture of innovation. The more innovation, the more lives saved and the more experiences improved.

I am not sure that I am the best person to write on innovation, but the idea of helping fill a void in healthcare technology today was something I wanted to have a part in. I was humbled when I was approached by the Healthcare Information Management Systems Society (HIMSS) in 2014, asking if I would take the lead on this important work. Always up for a challenge and passionate on the topic, I agreed. I don't believe anyone needs to be convinced of the acute need for increased innovation in healthcare technology. The gap may be closing, but other industries have certainly benefited far more than healthcare when it comes to innovation. This book aims to accelerate the closing of the gap.

I have several motivations to see innovation become part of our health-care technology DNA. The three that stand out the most are highly personal. My mother was diagnosed with stage 4 ovarian cancer in 2002, and after a valiant fight, she died in 2006. Along the way I was engaged in my mom's healthcare as she was shuffled from provider to provider, who all relied on paper charts. I will stay out of the details, but the opportunities for disruption were ever present. I always knew of the dysfunction of our industry, but now it was personal. I vowed to make a difference wherever I served to help save lives. My mother continues to motivate me today.

In 2012, after summiting Kilimanjaro with Texas Health Resources colleagues, we traversed to a remote Masai village in the great plains of Tanzania. Several months prior, we had made arrangements with the government to fund, build and staff the first ever medical clinic. We opened on time and to great fanfare. Our first patient was a young late stage pregnant mother who had not felt her baby in two days. We were not equipped for primary care, let alone trying to deliver a baby in severe distress. Our team of doctors were amazing. They innovated. They created all the tools needed out of our very basic supplies to including a sandwich bag and shoelaces. The baby was delivered still and brought back to life thanks to the innovation of our providers.

In April 2017, I was completing the national championships to secure my fifth consecutive spot on TeamUSA Duathlon. In the last two miles, I suffered a complete blockage of the left anterior descending coronary artery. I was suffering a heart attack commonly referred to as the "widow maker." In this case, innovation saved my life. When I crossed the finish line, I checked myself into the medical tent. The physician did an EKG via smartphone, and the strip was read immediately by the interventional cardiologist on staff at a nearby hospital. By the time I arrived via ambulance, the interventionist shared that Cleveland Clinic heart specialists had already conferred with him concerning the treatment protocol. Two days later I was back at work and anxious to heal quickly to race again. Using multiple Bluetooth technologies, all related vitals were transferred to my record and read by my care team daily. As a result, they were able to adjust meds in real-time, no appointment required. This hastened my recovery. Exactly 90 days later, I traveled to Denmark to compete in the Duathlon World Championships as a member of TeamUSA. As I crossed the finish line, arms outstretched carrying our nation's flag, I gave thanks for my life and all the people and tech that helped me not just survive, but thrive. This scenario would have played out much differently even one year ago. Thankfully there are some companies and organizations innovating. Now we need everyone to join in.

While there are many books written on innovation, including some titles related to healthcare technology, they largely remain theoretical. This is great as we need to proliferate the concepts and need for increased innovation. What is missing is a practical "how-to" guide. Using an established and respected innovation process framework, we wanted to share real stories of how individuals and companies are leveraging those concepts to realize innovation. Theory and frameworks are critical, but real-world examples bring them to life.

Voices leverages the HIMSS innovation framework and supplements each process with a handful of practical stories from people like you and me. They represent innovation taking place around the globe and in all aspects of healthcare to include providers and suppliers. It is my hope that you will find *Voices* to be practical and inspirational. It is designed so that any person or organization can take this framework and, by learning from others' experiences, bring innovation into your world. If everyone adopts an innovation mindset and way of working, imagine the stories of disruptive transformation we will share with one another as we seek to fulfill the promise of information technology in healthcare (Figure I.1).

Figure I.1   Innovation Pathways Factors of Innovation (HIMSS)

Source: Edward W. Marx, August 2018.

Chapter 1

Blend Cultures

Include the organization's larger community and ensure that institutional leaders are engaged and supportive of the proposed innovative strategies.

One of the first things you need to do to improve the odds of innovation success is ensure support and engagement from key organizational leaders. Innovation is hard to do in a vacuum. Often it takes a team, some of whom are directly involved and others who provide resources and political cover. As you embark on your innovation, take inventory of key decision makers, influencers and culture. Identify both the individuals who will help you and those who might hurt you. The more organizational community and leadership engagement you develop, the higher the likelihood of overcoming the obstacles that will be on your path to innovation.

* * *

Three Key Practices for Blending Cultures

Anna Pannier

Fourteen years ago, a CIO I worked with thought he was complimenting me by saying that I am "a bulldog in following up on outstanding issues." This image of a bulldog still looms large whenever I find myself in charge of a major initiative, especially one where I am having to be courageous and assertive. I hated this description of me for a long time.

However, I later came to realize that, yes, being a bulldog is a useful mindset for promoting change, innovation, adoption, and sustainment.

DOI: 10.4324/9781003372608-1

I simply need to be aware of my mindset and aware and welcoming of other mindsets that are critical in unison for success.

In the HIMSS Innovation model, blending cultures means including the organization's larger community to ensure that institutional leaders are engaged and supportive of the proposed innovative strategies.

As I reflect on my nearly 30 years in healthcare, three key practices stand out to me as the most valuable for ensuring engagement from the larger community and leaders:

- *Zoom out and zoom in*
- *Listen to understand*
- *Intentionally communicate*

I began my healthcare career with what I feel was an ideal job for a new grad entering the healthcare field: I was a business analyst for a large integrated health management organization (HMO), and I had the opportunity to participate in a long-term strategic planning effort. While I was just a new college grad, I had a seat at the table listening to amazing thought leaders imagine how healthcare will change over the next 25 years. This was the late 1990s, so our technology trends included beepers, headphones for our mini disc players, the Palm Pilot®, bulky car phones, and floppy discs.

The team structured the work into a SWOT analysis for our organization, identifying our strengths and weaknesses as well as the external opportunities and threats. Widening our lens gave us many insights into where the organization needed to grow and change in order to be successful 25 years into the future. It was very exciting to imagine mapping the genome to create customized medicine, increased cost burdens for acute care and medications, trends in an aging population and expectations of customized experiences (we called it mass customization then), growing specialization in medicine, medical tourism, and telemedicine, to name a few.

In addition to the wide lens for our macro view, we also used a close-up lens to get deeper into a micro view. The team included a fabulous PhD who was well-versed in statistics, survey tools, and analytics. Thanks to her expertise, we gained valuable insights from focus groups and staff surveys. We had the opportunity to meet with physicians and caregivers and shadow them to see first-hand how they were doing things today, their challenges, and their ideas for improvements and scalability.

While I did not remain at this organization for more than a couple of years, I learned the importance of managing my perspectives to ensure

I adopt a wide lens to look at the broad landscape and also zero in on the close-up.

Zoom Out—Zoom In

The concept of zooming out and zooming in was discussed by Rosabeth Moss Kanter in *Harvard Business Review* in March 2011. Zooming out provides us with a bird's eye view of the landscape, while zooming in completes the picture with the worm's eye view.

In terms of our innovation work, *zooming out* means understanding the big picture organizational strategy that helps ensure that our initiative has clear alignment. It allows us to understand the pillars or organizational goals that serve as a compass for clarifying context and what matters most. While all organizations have similar published priorities—patient satisfaction, safety, high associate and provider engagement, financial growth, reduced costs, etc.—meeting with executives and understanding where the organization is focusing its attention in the present moment is critical for informing any innovation effort. Innovation needs to stay aligned with the organizational priorities, which may shift and be more fluid than we have seen in the past. As we saw over the past couple of years with the COVID-19 pandemic, circumstances change, and so strategies must change to stay aligned.

Meanwhile, *zooming in* requires becoming immersed in the process and meeting with people where the work is done. Decisions about solutions made in the silos of conference calls, conference tables, work cubes, or home offices often lead to poorly understood processes and solving the wrong problems. Zooming in includes walking the process, mapping it, mapping the data in detail, and understanding who does what and when. Zooming in follows the model "Go slow to go fast" . . . just like the worm.

If zooming in and out represents our two eyes, then listening to understand reflects our two ears. As Epictetus noted, "We have two ears and one mouth so that we can listen twice as much as we speak." As a self-accepted "bulldog," listening—truly listening—requires extra effort for me. I have worked at this for decades (and always will) and now embrace that practice as *listening to understand*.

Listen to Understand

One example stands out when several years ago, I received a panicked phone call from one of my hospital CEOs voicing concerns about an upcoming go-live. This innovation project was aimed at transforming the physician EMR experience and had been a labor of love for more than a year. From a tactical perspective, we had a very comprehensive approach for communication and adoption and intended to roll it out in waves at different hospitals in the health system, with the first wave and first hospital—the one this CEO led—in just two weeks. My CEO was very concerned because his chief medical officer (CMO) was concerned. The words used could have taken me down a rabbit hole: "Anna, if this fails, it's on you. I am not owning this failure." But I listened to understand, and I realized he was scared of failure and needed his CMO to be supported and remain fully supportive and satisfied.

My next call was to the CMO, again to listen. This was the first major initiative that the CMO was leading. They had planned on having a smaller group of physicians that would be power users to help with adoption, but this group had not had enough time to feel mastery of the new tool. Under their leadership, the medical staff had also been the most aggressive in defining a date by which older practices using paper and dictation would be retired, so their activation had additional implications. The CMO was afraid of failure, both for themselves and for their peers. As the first of four waves, fear among one of our strongest physician champions, our go-live timelines appeared to be in jeopardy.

Through asking questions, we came up with a solution and clarified expectations. From a support perspective, I was able to explain the comprehensive support model we had in place from 6:00 A.M. to 6:00 P.M. daily. That provided some assurance, but the proof would be in the pudding once the medical staff saw it for themselves. The next thing we needed to address was understanding expectations for adoption and the four waves.

We agreed to alter the plan just a bit. Now, the first wave would be considered a *soft* go-live, and we would add a *fifth* wave to come back around to the first sight to shore up any challenges and have the formal final activation. This shift in expectation from a hard go-live to a soft go-live made an enormous difference to the way the CMO felt about the project, and it helped the physicians feel that they would have adequate support over a longer period of time, increasing the likelihood of success. This was especially critical because of the hard line in the sand that they themselves had made.

This project ultimately is one of maybe a half dozen of my favorite projects. It was full of innovation but also was a victory for me personally in understanding and improving my listening skills and the rewards of those improvements.

As we must adjust our lens to stay aligned and gain maximum engagement, listening to understand is a practice that, while challenging, is worth the effort. At a fundamental level, listening to understand is founded on honoring the person you are with. It also embraces a mindset of openness, inquiry, and caring.

It's important to bear in mind two things when practicing this style of listening: (1) Slow down and be present, and (2) embrace silence. When individuals are speaking, in addition to listening to the words they are using, try to reflect on what you are learning about them, how they are feeling, and what is important to them. By listening openly, you will understand that person's style of communication and can also learn from what they do not say. When asking questions, try starting queries more often with "what," "which," and "given" because these starters create more open space in a conversation and provide more insights.

Ideation—the sharing of ideas—can be an exciting time for some and a terrifying time for others. It is, therefore, helpful to establish clarity around the process you are undertaking. For example, if you are facilitating a brainstorming session, successful sessions follow the "yes, and" improv model to keep ideas flowing and avoid criticizing, which can halt the flow. Draw everyone out into the conversation, paying attention to who has not spoken up. Understand that some people will need more time than others, so allow uncomfortable pauses that encourage the less vocal to share. Some teams use walls or electronic boards to gather ideas over time, which can be a great way to engage broader participation and different shifts of people.

Listening to understand is a *practice*, meaning it takes life-long practice to continue to hone the skill. While listening is a component of communication, the final practice that I have found critical for ensuring broad and deep levels of engagement is *intentional communication*.

Years ago, one of my leaders shared this adage attributed to Woodrow Wilson: "Be brief, be brilliant, be gone." This is sage advice that, combined with specific successful techniques, delivers high levels of engagement.

As with listening, intentional communication has been and continues to be a challenging practice—but one well worth the effort.

Intentionally Communicate

Having ongoing feedback loops and clear communication and messaging for the broader organization relative to any innovation effort is essential. Culture is deeply embedded in the identity of an organization, department, role, and/or profession and shows up in the words we use. Sometimes, innovations may intersect with the identity of a group, and without understanding and acknowledging this, tremendous challenges can arise.

Not long ago my organization shifted from sole commitment to being a full stack development shop to one that embraced *low code* frameworks. The term low code is a terrible one to the ears of developers because it has negative connotations. Low code solutions are object-oriented development and lead to the growth of citizen developers that work hand-in-hand with esteemed programmers. Traditional developers are creatives, and the introduction of a low code framework was an afront to their creative identity. Over a few months, we had major upheaval among the developers. For some, this was an exciting new opportunity, while for others, this was the end of their journey with us as they sought to find another place to exercise their talents that allowed them to stay within their established culture and creative role definition.

Individuals often do not feel fully informed about changes unless they are engaged and hear about them multiple times. Who is talking about the change when trying to get individuals onboard also makes a difference. Corporate level announcements are one route, but often are interpreted as impersonal. Meanwhile, hearing about it within one's team, from a peer, or from a supervisor makes it more real and personalized. Hearing about things in a positive light from a person that one identifies with creates a tremendously positive impact. Producing videos showing new processes enables people to imagine themselves in the new process and is an easy way to ensure everyone gets the same message.

In *Switch: How to Change Things When Change Is Hard*, Chip Heath and Dan Heath introduce some valuable concepts about successfully leading change and innovation. One concept to aid intentional communication is the elephant and the rider as part of a communication model called the destination postcard.

The rider is atop the elephant and has very logical reasoning and can know exactly where they want the elephant to go. However, the elephant has its own brain and ideas, and it won't be going in the direction the rider wants unless the path is obvious, clear, safe, and desirable.

The *destination postcard* helps us curate our messaging to the space of a postcard and requires that we create messaging that appeals to both the rider and the elephant and makes the future destination clear, safe, and desirable.

People are inundated with emails and messages, so crisp communication lets people know the following:

1. What is new
2. Why it matters
3. Where to go for more information or questions

Some people really do want all of the details, and it can be common practice to attempt to answer every question ahead of time in messaging. Instead, make a safe space for the eternally curious and create transparency using linked frequently asked questions (FAQs). And always, always, avoid jargon because it makes people feel uncomfortable and like an outsider. No one wants to feel left behind, so our inclusive and intentional communication can aid in gaining the broad and deep engagement we need from all parts of the organization.

Another critical tool for intentional communication is stakeholder analysis. When I teach physician leadership with colleagues from the Tennessee Chapters of the American College of Healthcare Executives, we include this as part of a Change Management module. Every time we teach a new cohort, we see every participant finally realizing why they have experienced challenges in trying to create change—it's an eye-opening tool.

A great stakeholder analysis template includes the following:

▪ Name | Title | Department | Personal notes | Personal win | Preferred communication method
▪ Influencers (who can influence this stakeholder)
▪ Impact level (direct impact, affected, decision-maker, influencer, no impact, gatekeeper)
▪ Change status (strongly against, moderately against, neutral, modestly supportive, strongly supportive)
▪ Conflict approach (avoiding, accommodating, compromising, competing, collaborating)
▪ Expectations | Reason for resistance | Actions to address stakeholders | Assigned contact

Some aspects of this may be uncomfortable initially. However, it is exactly the type of preparatory work that ensures engagement and alignment.

Relative to any resistance, it can also be helpful to recall the five main reasons we resist change:

1. Fear we will lose something valuable.
2. Fear we won't be able to adapt.
3. We don't have enough information yet.
4. We don't trust the people implementing it.
5. Scars from negative past experiences.

Understanding these sources of resistance can help inform our approach, prevent teams from demonizing resisters, and lessen resistance. Resistance is actually a gift because it reflects a helpful human trait and serves as an innovation design constraint to work with. It can also turn a resister into a champion.

As you continue on your innovation journey, don't ignore the need to blend cultures and the three key practices to engagement:

1. Zoom in and zoom out to ensure alignment with corporate objectives and the real solutions.
2. Listen to understand to ensure engagement and deeper collaborations.
3. Intentionally communicate to ensure the highest levels of adoption and inclusivity.

Good luck on your innovation journey!

* * * *

Using Blended Cultures for Healthcare Innovation

Mark Waugh & Sergio Villegas

Murata Electronics North America, Inc. (MENA)

The cost of U.S. healthcare has risen at a staggering rate, nearly doubling between 2001 and 2015. In 2016, it accounted for *17.9 percent* of the U.S. gross domestic product. By 2030, the health spending share of GDP is projected to reach 19.6 percent. Factor in a pandemic or other global health crisis that drastically changes the landscape, and these rising costs are unsustainable. With new technological innovations, these costs can be lowered and can change the old health care model to a value-driven rather than volume-driven system. Value-driven innovation delivers solutions for big problems affecting many people more quickly and efficiently at a lower cost.

A substantial number of innovations in the healthcare domain come from multi-national companies and are targeted at world markets. These innovations can't happen without excellent, continuous communication to effectively manage the demands of all stakeholders. Thus, the ability of a company to blend cultures, that is, hold multiple frames of reference within itself, is a key component of taking a great healthcare technology idea from inception to realization.

Murata, Healthcare, and Innovation

Murata Manufacturing Co., Ltd. is listed by *Forbes* as one of the world's top multinational companies, with annual sales of more than $16 billion as of March 2022. The Japan-based company, started in 1944, integrates established technologies into the transformative world of healthcare innovation and bridges the divide between a large company and small start-ups—while addressing global cross-cultural business dynamics. With more than 80 percent of Murata's business outside of Japan, we must understand what is happening in the rest of the world—and how they do it.

The company is a worldwide leader in the design, manufacture, and sale of ceramic-based passive electronic components and solutions, communication modules, and power supply modules. Murata's healthcare technology is derived from our core technology strengths and experience in various fields, including PCs, smartphones, home appliances, and car electronics. Healthcare technologies produced by Murata include patient-monitoring applications, diagnostic and treatment equipment, medical sensors, medical RFID solutions, product distribution and control systems, and pressure measurement applications.

Healthcare technology is a moving target, especially in a post-COVID environment with many initiatives to drive remote healthcare monitoring in full force. In particular, new sensor technologies are poised to dramatically change the world's healthcare model. Wearable health trackers, for example, designed to help individuals improve their health and fitness levels, have been found to detect atrial fibrillation, a common heart abnormality that can lead to stroke. Thus, new technologies are moving the healthcare sector into a preventative and early treatment mode rather than treatment after the fact.

As an established company, Murata's expertise, resources, and ability to scale make it well positioned to take a great idea from incubation to market. Over the last several years, this model has not only been honed but also proven to be successful.

Shaze, Innovation, and Blending Cultures

The concept of *Shaze*, which loosely translates to company philosophy, is a uniquely Japanese concept. More than a poster on a wall, Shaze embodies a company's core values and establishes the mindset each employee should take in their daily activities. It's a key part of Murata's innovation culture.

Murata's philosophy, fundamental mission, and values, created by the founder Akira Murata in 1954, remain the same: To contribute to the advancement of society by enhancing technologies and skills; applying a scientific approach; creating innovative products and solutions; being trustworthy; and, together with stakeholders, being thankful for increases in prosperity.

Each Murata employee, whether in Japan, the United States, or another global location, is encouraged to exert "innovativeness" in their approach to all tasks, to come up with new and more efficient operational and administrative processes, and to help the company achieve optimum performance. Everyone is seen as contributing to the financial success of the company, its innovativeness, and the company's mission to advance society. Shaze extends to Murata's approach to collaborating with other companies.

Blending Japanese and American Business Cultures

The differences in business culture between Japan and the United States is so different that, without study and consideration of the other's culture, miscommunication and conflict is inevitable. Communications are subtle and nuanced in Japan, with an expectation that the listener will be reading between the lines. Japanese are more likely to rely on nonverbal cues and the context of what they say than the literal meaning of the words they use to say it. American business communications are quite the opposite and articulate intent directly.

In Japan, decisions are reached by consensus (which can take a while). Companies in the United States take more of a top-down approach—the boss or executives make a decision on behalf of the company.

Even the concepts of taking new ideas to the market are different. Murata looks at how its decisions and methodology will impact the company and its environment five to ten years out. It's a long-term innovation view, not new for the sake of new, and not new for right now. It's creating and introducing new technology to improve and advance society. Further, Murata executives see innovation as a journey involving a constant exchange of ideas and raising awareness of the goal and focus. To be an "innovator" within Murata also means considering how we may be able to improve our society and the environment and acting when these opportunities arise.

Blending Big and Small Company Cultures

While innovative, Murata, like many Japanese companies, is prudent in its due-diligence process. So, when it comes to working with a start-up, Murata is very diligent and organized in its process, with a focus on how we will work with a company, what we will each bring to the table, and the results we would like to see. We want to build long-term, viable, profitable, sustainable solutions and products. We like to be extremely thorough when evaluating opportunities.

It's a different approach than that of a start-up, which tends to be more action-oriented than planning-oriented. Start-ups are agile and make decisions quickly, often taking greater risks than would a large, established company such as Murata. But bringing in that careful planning mentality to a start-up is helpful for execution of an idea—even if that path changes. The careful planning approach also shapes how we track results. Start-ups want to see an immediate relationship between effort and results. Japanese companies, and Murata in particular, look at innovation as more of a journey.

To bridge these cultural gaps—between American start-ups and a Japanese global company—Murata Americas collaborates with start-ups through our MENA Corporate Technology & Innovation (CT&I) group. The mission: Accelerate innovation and develop new technology through industry partnerships and by leveraging Murata's corporate strengths (Figure 1.1).

The group's key strategic focus areas are research and development, technology scouting, business realization, and product engineering. By creating an environment where all viewpoints are respected and innovation is celebrated, this helps ensure that business goals are aligned. Further, the group is agile enough to harness emerging market opportunities, such as MedTech, CleanTech, and DeepTech (Figure 1.2).

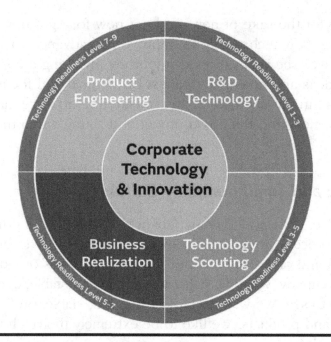

Figure 1.1   MENA CT&I key strategic focus areas.

Figure 1.2   MENA Open Innovation Event—themes of interest.

CASE STUDY: OPEN INNOVATION EVENT

Corporate innovation is exciting and can reap large rewards for an organization, ensuring that it keeps a pulse on trends, maintains a competitive advantage, and seeks parallel opportunities synergistic with the company's core assets. However, innovation is not easy, and often in large organizations, multiple layers of managerial bureaucracies exist that can derail or hinder the executional speed necessary to compete in an ever-evolving global world. For some corporations that are newer to implementing innovation initiatives (e.g., incubators, accelerators, CVCs), there are many best practices that companies can leverage to kick-start and ignite new ideas both internally and externally.

One such practice that Murata initiated in 2020 was to launch an annual Open Innovation Event to attract early-stage start-ups with a focus in MedTech, CleanTech, and DeepTech. The key goal of hosting the pitch event is to identify potential engagements between the start-ups and Murata (e.g., JV, POC, investment, licensing) and hold preliminary discussions between the two parties that often lead to deeper exploratory conversations.

Over the last few years of hosting Murata's Open Innovation Event, the team has reviewed hundreds of start-ups and learned a lot along the way, namely the following:

- Likely true for most technology-led organizations, it is critical that the screening metrics be aligned with technology wish-lists/gaps and long-term strategy of the various product and R&D teams and that key stakeholders from those groups be involved through the screening and voting process.
- Try to find an engagement solution that will benefit both parties. While most start-ups are looking for a financial award, many are also looking for strategic partners that can provide open access to the supply chain, design for manufacturing assistance, new customers, marketing and sales support, and/or technical feedback (we call this Smart Money) that allows the start-up to validate its business plan and technology feasibility.
- Ensure that there is an owner of the relationship with the start-up who will be the internal champion of the engagement. After the pitch event is over, the collaboration is truly just beginning, and the internal

champion will need to work with that start-up to execute the proposed engagement idea.

■ It's critical to understand budgeting and the flow of funds within the organization to pay for engagements with external start-ups. In a large corporation, the main stakeholder and champion isn't always the individual or team providing the necessary funding. Therefore, Open Innovation staff and engagement champions need to work in parallel to promote discussions internally about funding and justify why the engagement is important.

Open Innovation Events take a significant level of effort and are a fantastic innovation tool that companies can leverage to open the top of the funnel for new business ideas and collaboration. While it is logical for a large organization to focus on its core business, unfortunately it is also common for large organizations to lose sight of new innovations that can disrupt their core and future business when its teams are hyper-focused on existing business, its R&D pipeline, and its five-year strategic product roadmap. Additionally, there is often a healthy sense of optimism within R&D and engineering groups that they can do everything themselves, including maintaining existing research and products, creating new solutions and products, and keeping a pulse on the market. Accomplishing this large order requires a dedicated team to manage such efforts.

Murata has realized that to continue to create new tangential business and maintain its competitive position as a global supplier of electronics, it needs a pipeline of new external ideas and ways to work with other companies, start-ups, universities, and thought leaders. Murata's Open Innovation Event is simply one tool in its toolbelt used to engage with start-ups and create new ideas, products, proof of concepts, and joint ventures. We would like to explore areas where we can work with our partners to combine each other's existing core technologies and create value for existing and new customers.

Takeaways

1. **Don't underestimate the importance of and challenges with culture blending. Cultural blending is a key first step for companies to successfully work together.** Mutual respect is essential for

successful operations, and there must be a personality match between corporate cultures for joint ventures to succeed. Acknowledging potential cultural issues early and educating employees on such dynamics allows them to appreciate differences that can help build a stronger organization in the long run.

2. **Keep in mind that both parties can learn from each other**. For example, to work with a large company, a small company needs to understand what drives the larger company, and big companies need to remember they were a small company once. There are different challenges at a start-up level, such as restrictions on money and time. At a start-up, there is typically a small group of high-performing people, whereas at a large company there is a large group of diverse capabilities.

3. **Culture blending isn't quick, and trying to force it doesn't work.** In the case of healthcare innovation, some projects take years to bring a great idea to market. But many start-ups want quick gratification and returns on investment. These types of projects might not be successful because such a mindset might not understand a slower market and development strategy. In the United States, healthcare equals big money and big potential, whereas the driving forces in other geographies are different.

4. **While you may have a preconceived vision in your mind about the other company, such as what they do and how they operate, you won't really see and understand each other until you personally meet and interact with each other**. Japanese companies such as Murata are very interested in a company's background, to understand its business. In Japan, with its high context culture, a successful collaboration is as much about the people as the product or technology.

5. **Keep in mind how you do things at your company and in your country may not always be well-perceived by others. It's important to have "ambassadors" to bridge cultural divides to translate and educate the other company on your company's operation style**. For example, sometimes a Japanese company's (e.g., Murata's) driving need to understand the other company and its value proposition can be perceived negatively, as being doubting. But it's not. It's a method to achieve as much understanding about the other company as possible. We ask many questions on our fact-finding missions, and it's a careful process. By contrast, in the United States, many

collaborative ventures are based on "gut feelings" and limited, but direct, questioning. Understanding another culture's business etiquette is important to closing deals.

6. **Remember, it's an innovation journey. Innovation itself is a process, and taking the long view increases your chance for success.**

7. **Measure twice, cut once. We believe at Murata that more due diligence up front leads to less confusion and better understanding later.** Of course, questions need to be asked in a way that does not convey disrespect. Even disagreements can foster mutual understanding through discussing and understanding the issues.

8. **It's not just communication between the companies' executives—internal communications are important too.** Consistent and clear communication with internal stakeholders strengthens the mix, the blend if you will, of the ideas and insights coming in from all members of the collaboration.

9. **Identify an internal champion.** Ensure that the organization has a relationship owner with the start-up who will be the internal champion of the engagement. After the pitch event is over, the collaboration is truly just beginning, and the internal champion will need to work with that start-up to execute the proposed engagement idea.

10. **Enlist an engagement solution that aligns with and benefits both parties.** While most start-ups are looking for a financial award, many are also looking for strategic partners who can open access to supply chain, design for manufacturing assistance, new customers, marketing and sales support, and technical feedback (we call this Smart Money) that allows the start-up to validate its business plan and technology feasibility.

Conclusion

Innovation thrives when different knowledge domains come together. Many discoveries have been born following exposure to indirectly related ideas via other companies, start-ups, universities, and thought leaders. Similarly, developing relationships is critical to building sustainable achievement. While subjective, cultivating relationships needs to be pursued with equal discipline to pursuing technological innovation. Only by blending cultures can companies leverage each other's core competencies and technologies. World-class organizations understand that great innovation is achieved only through aligned goals, mutual trust in each other's competencies, and a commitment to open and direct communications.

Blending a Revenue Cycle Culture

Charles Christian
Indiana Health Information Exchange

After a long implementation process to replace our revenue cycle applications, I was walking in from the parking lot with our CEO, a long-time friend. We were talking about how long the implementation had taken and the impact on the organization. I was thinking that my team and I might be getting a few kudos from the boss, but that is not what I heard next. In so many words, the CEO mentioned that he had seen our days in accounts receivables climb at a growing rate, and he was being told that all signs pointed to the new revenue cycle system; he wanted to have further discussions about what it would take to reimplement the system we just pulled out. Needless to say, the walk in from the parking lot just got a little longer.

Our organization resided in a community that was one of the earliest territorial settlements, therefore, our culture was founded on hard work and trust. We trusted that everyone was working hard to keep the organization successful, which was the case, but as I learned, the assumptions need to be revisited from time to time.

Over the course of the next many days, there were lots of conversations around seeking answers and hard data to back them up. These conversations included the CFO, director of patient accounting, and internal audit. Unfortunately, there didn't appear to be agreement on the underlying issues; falling back on what we believed, rather than what we know.

Working with internal audit, we found that we needed to move from a culture of "what we think" to one of "what we know." Working together, we quickly completed a detailed analysis of the current state of our accounts receivable (patient claims being processed) and historical data. With this review in hand, we only had a portion of what we needed to work toward the cultural shift. We needed to look at the proverbial "three-legged stool": People, process, and technology. Right now, we were only able to see one side of that triangle.

As part of another cultural shift, the organization was also moving to embrace the tools of continuous quality improvement. Applying these to this issue would be a challenge, but the tools did fit the task at hand.

A cultural shift, especially one of this nature, is not something for which you can stand up in the front of the boat (room) and point the way. The people doing the everyday tasks need the opportunity to engage, understand, unlearn the old ways, and embrace/learn the new. We embarked on a journey that would allow management to learn from their teams; it was a learning experience for everyone.

Over several weeks, each member of each team had the opportunity to map out their current processes with sticky notes, which would eventually cover an entire wall in our conference center. Complex and disjointed don't come close to describing the final resulting "process map," if you could call it that.

Something unexpected occurred during the "mapping" process; those watching their teams describe how the process currently works (for them) realized that each person with the exact same job duties was going about their jobs differently, with different outcomes. The teams found (for themselves) that the culture of "this is the way we've always done it" just wasn't working, and it needed to change for the organization to become successful.

I observed that you can't underestimate the power of allowing people to discover, on their own, something that you may already know to be true. It's a teaching moment that creates ownership in the change or shift; an ownership that needs to be taken, rather than given. Did I mention this was a learning process for everyone?

Armed with an understanding of the present, the teams moved forward with creating the future, the shift to new processes, and a culture based on standard processes, metrics/measures, and accountability; a culture of what we know rather than what we think.

The next question was what to do with this new knowledge, and how would the culture be incorporated into the fabric of the everyday. Having addressed the "process" leg of the stool, it was time to add the "people" leg to the stool.

During the process mapping, the teams expressed a concern about their level of knowledge regarding the new revenue cycle application. Management had determined that if they only trained their team members on the specific functions that they thought they would need to effectively do their jobs, it would decrease the total amount of non-productive time set aside for training. In retrospect, this decision had created "process blinders" for their teams; they couldn't see the upstream or down-stream processes, making them blind to the previous and next steps in the very complex process of claims processing. The culture of "what we think" was again at work.

To address this knowledge gap, every member of the claims management team was provided with a comprehensive re-education on all the features/

functions of the new revenue cycle solution. Armed with a new level of understanding, the team asked for some time to rework the previously redesigned workflows and processes. The culture of "what we know" was taking root.

Before I add the last leg of the stool, "technology," I need to provide a view of the physical environment that housed the revenue cycle teams. Picture a large, open space with rows of desks and large open filing cabinets lining the walls around the perimeter—filing cabinets that held the 70,000 to 80,000 open/active accounts (paper records) in various stages of the billing process.

The next step in the cultural shift involved the "technology" leg of the stool, but it also had an associated impact on the "people" leg as well. If we were to leverage the new processes and feature/function of the new revenue cycle solution, we needed to do something about the tsunami of paper that rushed off the printers each day, requiring filing, which delayed the claim submission and follow-up functions. Document/image management was the break waters that we implemented to calm the on-rushing paper waves. The desire was to move from "where did we file that," to "it's here with a couple of clicks of the mouse."

To root the cultural change further into the organization, management also redesigned the physical workspace of the revenue cycle team, moving from a large open space with thousands of file folders to a newly painted and carpeted area that housed individual cubicle workspaces and sound dampening surfaces to create a quiet environment that was pleasant to work in the many hours of the day.

Moving from a culture of "what we think" to one of "what we know" required the development of measures and metrics that allowed us to "know." With any quality improvement process worth its salt, there must be a "check phase" where the assumptions are reviewed and tested, adjustments made, and progress tracked. Do we have strong roots that run deep in the organization like those of an oak tree, or are they superficial and easily removed like those of a corn stalk?

The outcomes from this effort tracked as we had hoped; our days in AR moved in the correct direction, the percentage of claims that had to be resubmitted declined, immediate cost savings occurred with the reduction of filing staff (they were re-tasked to the follow-up team, relieving the need for recruitment), staff turnover declined, while staff morale greatly improved. We declared it a success and continued to monitor progress so that we would "know what we should know."

Finally, the lessons learned were many. First, an issue is probably different than what you think. The real underlying cause or causes require a look at all aspects, especially the culture of the area/department/organization. Second, the work to make modifications to any culture requires a commitment from leadership, good data, partnership with all parties, and a willingness to create the measures and metrics to ensure that you're growing good cultural roots. You have to measure what matters and leverage data to combat emotions that typically come when you try to blend cultures.

Balancing Innovation and Institution

Pete O'Neill

Innovation is one of the pillars of Cleveland Clinic; it is part of our culture. Our hospital was created by doctors returning from World War I who wanted to establish a hospital to deliver the best patient care, and over the next almost-100 years, Cleveland Clinic has pioneered incredible healthcare "firsts." Our leaders are innovative people with a history of delivering innovative solutions to patient care and hospital operations. However, despite our demonstrated organizational and personal interests to be innovative, we still need to be rigorous about how we manage the delicate balance of promoting innovation while running a large, multinational hospital system, and protecting our most precious stakeholders: Our patients.

A culture of innovation requires that everyone in the organization feels empowered to open their minds, to identify opportunities to improve what they see around them, and to think about out-of-the-box ways of delivering patient care. At Cleveland Clinic, we consider all 55,000+ employees "caregivers"; all of us have the opportunity to improve the way our hospital delivers care to patients. We see innovations and insights into needs for innovation coming from all parts of our organization.

However, any effective and responsible organization needs processes to manage innovation. It would be bedlam if every one of 55,000 employees, or individual departments, or any other unit pursued their own healthcare innovation strategies. At the same time, every idea that comes from any part of the organization can't get the full consideration of our leaders. The trick is to find the right balance of nurturing innovation, which is an inherently creative process, with rigorous processes to achieve organizational efficiency.

It's useful to think about our approach to managing innovation as a pyramid. The broad bottom is our sources of innovation, with ideas coming from all parts of our organization. As each innovation is evaluated, it gets elevated through the organization (up the pyramid) based on its potential and probability of delivering impact to the organization. Once an innovation reaches its maximum level of impact, an appropriate group of Cleveland Clinic leaders reviews the innovation in the context of our corporate strategy and then decides if the innovation will be "approved" to be pushed down the pyramid to all the areas under the leadership of the decision-making group.

Of course, the most impactful innovations are the ones that can improve the entire enterprise, maybe even be scaled so they can be delivered outside our own organization. These innovations rise to the top of our pyramid and are reviewed by our most senior leaders. Some innovations don't have the potential to affect the broad enterprise; they should only rise to the level of their appropriate impact.

One challenge is determining which innovations have limited impact before they get to the highest level. Sometimes the potential for an innovation isn't known until it is shared with leaders with the most strategic perspectives; if these innovations are turned down too soon, their full potential will be missed.

Another challenge is making timely decisions about which innovations will be approved to be implemented. All innovations have a shelf-life, influenced by market conditions and the persistence of the source of the innovation; if every innovation is presented to our highest leaders, then it will clog and delay their decision-making processes.

To execute on this balance of determining which innovations to push to our highest level, we need to have efficient and trusting communication. We need to be comfortable pushing down as much authority and strategic understanding as possible. We need to recognize that innovation can be messy and risky, and being innovative means being accommodating to working on messy and risky things.

Cleveland Clinic has a wonderful spirit of collaboration among its leadership, and we're always looking for new ways to improve how we incorporate innovation. For example, today we're working on ways to bring more external innovations into Cleveland Clinic. Managing innovation is something an organization probably never gets completely figured out. Being truly innovative requires continuous reflection on and refinement of how innovation is managed, based on external and internal forces.

An organization that desires to have a rich culture of innovation must encourage and nurture all of its employees to be empowered to innovate. A *culture* of innovation can't be nozzled or regulated. We can't tell people we want them to be innovative according to a set of institutional rules. However, we need institutional rules and processes to convert raw innovations into actionable opportunities for the enterprise. Within healthcare, we have the additional balance of managing innovation according to the security requirements of treating patients. At Cleveland Clinic, we don't have all the answers, but in our nearly 100 years, we've embraced the challenges and look forward with enthusiasm to the opportunities that are in front of us.

Chapter 2

Use People with IT

Do not create an over-reliance on people or on technology; use both resources in concert.

Often we rely too heavily on technology as we embark on innovation. Sometimes innovation starts at the other extreme with people but little incorporation of automation or tools. The best innovations tend to be the result of a strong balance at the intersection of people and technology. Always take an inventory of people and technology to ensure balance. It is the ability to take the best of people and technology, then melding them together, that ignites innovation.

* * *

Roberta Schwartz & Bruce Brandes

The Existential Threat to Healthcare Organizations

If Abraham Maslow were creating a hierarchy of needs for acute health systems and long-term care facilities, an engaged workforce and an economically viable care model would be as existentially foundational as air and water.

Today, healthcare executives are consumed with three realities:

- There aren't sufficient caregivers in the United States to staff healthcare organizations.
- The work required for care of an increasingly acute population is causing caregiver strain and burnout.
- The current costs of care delivery are fiscally unsustainable.

The physical, emotional and economic burden on our caregivers has never been more challenging. Beyond the human toll, the resulting burnout and workforce shortages have exacerbated the already excessive costs of care. The challenge of staffing is causing escalating costs, which is unsustainable for healthcare organizations.

How can you improve quality, safety and compliance AND lower labor costs at the same time? Care organizations must reimagine the underlying business and clinical models of how care gets delivered with transformational change, not incremental improvements.

Learnings Proven in Other Industries

In transportation, with the addition of backup cameras in cars, we now have fewer parking lot fender benders and avoid looking over our shoulder when in reverse. More advanced capabilities, such as automatic braking, lane change assist and dynamic rerouting, are among the first features toward autonomously protecting against human frailties. The benefits have both true safety and cost implications to both property and human life.

Many other industries, such as space, manufacturing, retail, distribution and the military, have already proven the same technology of advanced sensors and artificial intelligence to accelerate innovation, maximize human capital and improve quality, safety and compliance.

In any FedEx distribution center, ambient sensors and data science have radically improved efficiency and quality while lowering costs. Employees neither "clock" in and out nor step away from their core work to type a document describing the task they just completed. In an Amazon Go store, shoppers are encouraged to enter the store, pick up their items and walk out, with simplicity that can only be achieved when technology becomes invisible.

Introducing Ambient Intelligence for Healthcare

Google defines ambient intelligence as technology that enables video and audio devices to interact with and respond appropriately to the humans in that environment. In healthcare, seamlessly embedding this technology within the fabric of care delivery becomes a strategic enabler for the staff to be targeted in the way that they provide care.

Thoughtfully designing and responsibly deploying the most valuable use cases require care teams to specifically define the following:

■ What is the event we are looking to capture?
■ What triggers the activation of monitoring for this event?
■ How and with whom should we communicate the alert (real-time vs. retrospective)?
■ What integration is appropriate with other systems?

When seeking examples in healthcare, one looks no further than stepping into the patient room where we check whether staff are washing their hands, whether the patients are making micromovements that may warn the staff about an impending fall, whether the patient has been turned in the last four hours, or whether the staff need education for appropriate lifting techniques. In the past, each of these would require human intervention for review and then intervention. Just as self-braking eliminated complete human intervention to stop a car, ambient intelligence in healthcare can eliminate human intervention to get information from the patient to the individual truly needed to create a care experience.

4 Key Pillars for Ambient Intelligence for Healthcare

To harmonize ambient intelligence with people who deliver and receive care, there are four pillars that must define this solution for healthcare.

1. *Omnipresent:* Ambiently and continuously collect data, looking and listening, 24x7.
   As humans are limited to only be physically in one place at a time, we can never have enough caregivers (due to caregiver shortage and cost) to always be present to determine what is needed by every patient at any given moment in time.
   Capturing all the clinical, operational and environmental data, critical to fully optimize care, is constrained by human limitations such as seeing in only two dimensions, selective listening and only being able to physically be in one place at a time. Further, too often we are limited by the subset of that information that the caregiver subsequently types or dictates into the EMR. Strategically placed advanced sensors comprehensively capture critical new data in real time. The smart sensor is not a camera and not

continuously recording them—but rather only logs defined events, identifying presence or absence of an event, and notifies in real-time or after-the-fact. The "smart" element of the sensors ensures no human monitoring is required on the other end of the device as information is coded to artificial intelligence and appropriately communicated to the caregiver immediately.

When the right information can get to the caregivers at the right time, intervention can occur more quickly and with less resource utilization. When oversight can occur that allows for targeted education rather than "everyone is educated because we aren't sure who needs it," it saves educator time and staff time.

At a time where resources are precious, *omnipresent* is a friend to the caregiver.

2. *Omniscient:* Intelligence to fully understand and dynamically apply information in real time.

Despite best intentions and efforts, human beings can be influenced by personal frailties and limitations, such as fatigue, stress, emotion, memory, implicit bias, experience, education, etc.

With a transformative amount of new data being captured, ambient intelligence automates continuous learning and improvement to optimize how, who, where and when to take action.

The vast amount of continuous data is constantly processed in real-time, with more complete context, to support caregiver decision-making.

The concept that staff are prompted when tasks need to be completed can lessen burden of the caregiver to remember each step.

As any concerns of artificial intelligence usurping clinical control gives way to the idea of ambient intelligence and in-the-moment decision support, clinicians can be secure in the confidence they are delivering the most appropriate, safest, knowledge-based care.

3. *Contemporaneous:* Ensure timely action, simplify workflow change and leverage existing investments.

With ambient intelligence creating increased clarity and focus on prioritized actions that people must take, we also need to eliminate friction. Limitations and inefficiency of human communication, such as physical proximity, language, handwriting, misinterpretation, and distraction, can all lead to delay or failure to act.

From workflow and integration perspectives, by seamlessly working with existing assets such as mobile devices, nurse call systems, electronic health records and patient monitors, operationalizing ambient intelligence requires minimal change management to immediately realize significant new benefits.

The hope that devices can collect, through voice and video, acts that are being completed by caregivers and directly populate the EMR without human intervention will reduce "pajama" time for caregivers and allow staff to leave on time and concentrate on home-life when at home.

4. *Compassionate:* Empower people to provide care with a more human touch.

By making powerful technology invisible, each caregiver currently burdened by inefficiency and distraction of unnecessary tasks is empowered to be physically and emotionally present where they are needed most.

Increasingly, our clinicians are being refreshed with the reminder of why they originally chose their profession—and spending more of their valuable time doing what humans alone can do (and what is needed most)—love, care, touch and heal people.

Practical and Compelling Initial Use Cases

With a focus on restoring the humanity of care delivery to reenergize our caregivers, there are immediate clinical and operational use cases every health system should consider.

Clinical:

- Falls and pressure injury prevention
- Infection prevention and control
- Elopement avoidance
- Incontinence intervention
- Virtual nursing

Operational:

- Rounding and regulatory compliance
- Environmental services high touch–surface cleaning
- Medication disposal
- Hand hygiene compliance
- Clinical documentation capture
- Surgical suite efficiency

Benchmarks for Success

While the use case possibilities begin to seem endless, the transformational promise of ambient intelligence in healthcare will be fulfilled by aligning efforts to current organizational priorities, optimizing workflows that best leverage the technology and quantifying targets and outcomes.

Initial benchmarks for success to measure include those shown in Figure 2.1.

The Future is NOW

For many legacy healthcare delivery organizations blunted by the human and financial cost of care, keeping the status quo or just tweaking around the edges of new care delivery models is no longer an option. New competitive pressures and economic realities have created existential risks.

At the same time, the power of ambient intelligence has been proven in many other industries to harmonize human and machine. Ambient intelligence in healthcare is a foundational solution to enable the transformational promise of the smart care facilities of the future to provide a level of care the world has never seen.

Categories	Use Cases
Performance	Length of Stay
	Catheter-Associated Urinary Tract Infections (CAUTI)
	Central Line-Associated Bloodstream Infection (CLABSI)
	Pressure Ulcers
	Falls
	HCAHPS scores (overall and each sub-category) - Overall Quality
	Sub-Cat 1 - Nurse Comm. Compos.
	Sub-Cat 2 - Responsiveness Compos.
	"Quality or other projects" completed by staff (i.e., measure of free time)
Utilization/Compliance	Sitter usage
	Documentation in additional areas such as malnutrition
	Number of texts to each family
	IPAD use by patients
	Time to document patient record by nurse
Satisfaction	Timeliness of call light answering
	Potential self-reported survey by staff of feelings of happiness with role; time; staffing
Staffing	Staff Engagement Score (annual)
	Staff retention - Turnover by years of service
	Ability to fill staff openings / Vacancy rate
	FTE dedicated to each function
	Number of auditors it takes or audit hours for various quality roles

Figure 2.1   Initial benchmarks for success to measure.

Support Digital Processes and Technologies with Governance

Danielle Wilson, Angela Skrzynski, Tarun Kapoor &
Krishna Kurapati

Necessity is said to be the mother of all invention. While the adage doesn't mention a global pandemic, most hospitals and health systems began innovating in early 2020 as COVID-19 patients overwhelmed facilities and operational decisions were made to defer elective surgeries, wellness visits, and non-critical tests and procedures.

But many non-COVID patients with emerging or chronic conditions still had to be seen. Follow-up visits still needed to occur, as did screenings for cancer and other life-altering conditions.

Like many hospitals and health systems, Virtua Health quickly ramped up its digital and telehealth efforts to reach more patients how they wanted to be reached—at home. But unlike many health systems, Virtua instituted an operational support department and governance model in the form of a Digital Transformation Office (DTO) to ensure that these new patient engagement models worked as advertised, garnered sufficient buy-in, benefited both patients and practitioners, and became ingrained in the organization's caring culture and model of service.

A recent study shows that 90% of patients who were unhappy with their care experience indicated they wouldn't return to the practice or recommend it to others.[1] Retaining happy patients is more cost-effective than recruiting and onboarding new patients, so any new initiatives must be undertaken carefully.

Since focusing on digital engagement strategies, Virtua has seen a 400% increase in patient usage of digital tools, and patients say the web-based digital solutions are easier to use and more accessible than previous efforts. Three of Virtua's new telehealth digital programs combined boast a net promoter score (NPS) of 75%—more than double the industry average of 31%.[2] Establishing new workflows, adopting new technologies, and transforming patient services require the right blend of people, technology, and governance to be successful at the health system level. Virtua Health's example can serve as a case study for other healthcare organizations wanting to create lasting change that benefits both patients and providers.

No Choice But to Go Digital

Virtua Health is a nonprofit regional health system that serves southern New Jersey and the Philadelphia area. The system includes five hospitals, seven emergency departments, eight urgent care centers, and more than 280 other locations. The health system staffs more than 14,000 people, including more than 2,800 physicians, physician assistants, and nurse practitioners.

As a nearly complete nationwide shutdown in healthcare commenced in March 2020, telehealth became a lifeline for patients who needed care and for hospitals and providers to see patients outside the facility walls.

Within days, Virtua Health adopted fully remote digital health practices to continue to treat patients. The health system had been moving toward creating more patient-friendly treatment options, as evidenced by its strategic priorities for 2020:

■ Transform and improve the delivery system.
■ Evolve and align the system's caring culture.
■ Orient offerings toward the consumer/patient.

In summer 2020, Virtua leadership took a step back and examined the workflows they had created to determine what was working, what wasn't, and what was required to continue on the innovation path.

They determined that patients increasingly wanted and expected a digital-first experience mirroring the types of service they receive in non-healthcare settings (e.g., banking, commerce, hospitality). For example, it was evident that high call volumes were resulting in lower service levels, introducing friction and frustration into care processes. Leaders also wanted to reduce in-person exposure to potentially infectious environments, recognizing the patient and family mix of inoculated, unvaccinated, and those with compromised immune systems.

Finally, the health system recognized that legacy technology wouldn't provide the type of care experience that patients were demanding. The system required new technologies to supplement in-person experiences and to handle rising volumes of patients in the ED and throughout the hospital.

Leaders pointed specifically to the existing telehealth technology, which was app-based and cumbersome to download and use for both patients and clinicians. Patients were put off by having to download an app, and the governance team believed a new platform would allow the health system to be more flexible delivering care in new and evolving clinical settings.

Opinions about the telehealth technology proved prescient, as a 2021 healthcare survey indicated that 75% consumers expect telehealth to remain an option for patient visits, and 83% want to keep self-serve options that gained traction during the pandemic, including online check-in, digital payments, and electronic communications with providers that include billing.[3]

From those discussions, the DTO was born to help facilitate needed technology and other investments, to align initiatives with organizational goals and culture, and to obtain the necessary buy-in from across the health system to create lasting change.

Governance Provides Necessary Structure for Success

While the DTO oversees digital strategies at Virtua, the office isn't overly bureaucratic. Rather, the office provides a governance structure where various stakeholders—including IT, clinicians, nurses, legal, operations, and those affected by the specific business process under discussion—come together to share ideas and tactics to move a project forward.

A member of the team often plays the role of patient/consumer, providing valuable input into why a program should be launched and how a patient would likely interact. The goal is to select the right digital technologies and practices, safeguard patient information, and innovate quickly while maintaining clinical integrity.

During initial discussions, an advisory committee quickly coalesced around the theme of "Get success and get there quickly," recognizing the required pace of change. Digital engagement comes in multiple forms, including "digital first" and "digital only," and both may require human interaction at some point, given that all digital technologies have their limitations.

Digital governance has its own structure, reporting to the strategic advisory team (i.e., C-suite). Because digital projects affect every corner of the health system, the digital transformation steering committee includes senior leadership from across the organization.

Once the steering committee has approved an initiative, operations staff are involved for planning, execution, and implementation.

Virtua takes an agile approach to innovation, launching projects that are 100% clinically relevant and sound while the technology and related processes may only be 80% to 90% completed. Perhaps the user interface needs a bit more work, or some patient features aren't fully developed. By launching earlier, staff can gather feedback about what's working while optimizing functionality issues and incorporating any feedback into future iterations.

Not every program will prove successful, but that's okay. The goal is to keep thinking about how digital technologies and workflows can transform interactions with patients. This includes trying new things, expanding those that resonate with patients and deliver value to the organization, adjusting those that don't meet expectations, and eliminating programs that aren't working.

Governance empowers Virtua leaders to innovate, putting organizational authority behind projects, supporting successful efforts, and obtaining critical buy-in.

Change Management Component Cannot Be Overlooked

A recent survey of health executives showed that 83% said improving patient access with digital tools is a top strategic priority.[4] The American healthcare system is also dealing with pent-up demand for medical appointments. A national survey published in February 2022 shows that nearly one-third of adults 50 and older postponed or canceled a healthcare appointment for pandemic-related reasons. In addition, 75% delayed primary care visits, and 72% put off tests and procedures. The survey also indicates that 22% had not yet rescheduled physician visits and 26% had not rescheduled tests or procedures.[5]

While digital tools have shown their value to connect with patients and enable care, signing a contract to bring new technology into the health system is the easy part. The hard part is changing minds to convince caregivers that the new technology will improve the patient experience and their own workflows.

Change can be difficult, and adopting new processes can be like trying to alter course on an ocean liner using only oars. Virtua digital team leaders say implementing new technology represents 15% of the total challenge, with the change management component encompassing the remaining 85%.

The key to success starts with finding those two or three champions who recognize the value of the change and its potential to transform processes. Virtua calls them "viral adopters." Often, these viral adopters can generate enthusiasm about the new technology or process, answer questions, and help assuage any fears or doubts.

Support from the C-suite comes into play for those who don't immediately buy in. Storytelling with data is crucial when describing the importance and impact of a project and how it supports the strategic imperatives to make Virtua the community's provider of choice. Consistency in messaging is critical to enact new processes and workflows. The overarching governance model provides the support necessary to effect change, bolstered by internal relationships and the common desire to continually improve.

Put the Right Metrics in Front of the Right People

Across projects, the DTO keeps track of more than 100 operational metrics, and every digital project has a dashboard that shows which aspects are working and which are not, including departmental impact metrics. But that doesn't mean that everyone across the organization needs to focus on every metric. While each metric has value, operational owners are the ones who monitor operational metrics, looking for trends, potential new projects, or possible improvements to existing programs.

Those 100-plus metrics roll up into key performance indicators (KPIs), or governance metrics, that are shared with senior executives and across the organization. Of those KPIs, only a subset is elevated into system level goals. The purpose is to closely track a few meaningful data points across the organization instead of trying to keep up with all of them, a mistake that many organizations make. Data overload creates confusion and competition among projects.

Governance puts a structure around projects and data points, helping organizations keep their focus on what's important.

A Series of Successes

Embracing the "Get success, and get there quickly" theme, Virtua rolled out several new programs using agile techniques and QliqSOFT's Quincy platform that are delivering significant outcomes for patients through virtual consultations, chatbots, and remote patient monitoring (RPM), while showing value to the health system.

Urgent Care Telehealth (UCTH)

Started in March 2020 and expanded in July 2020, the UCTH program conducts browser-based video visits 365 days a year. Conditions treated include COVID, upper respiratory infections, rashes, urinary tract infections, ear and eye complaints, musculoskeletal injuries, and more. Of the patients seen in urgent care, about 15% make return visits. Volumes have increased steadily since the start of 2021, with an uptick after April 2021, the month that they made the transition from an app-based platform to a web-based platform.

Urgent care visits from non-COVID-related conditions continue to climb. This increasing volume can be attributed to increased consumer awareness from informational campaigns, word of mouth, and ease of use of the tele-health program.

Elsewhere, a small-scale study shows that incremental changes to work-flows can improve organizational efficiency, allowing practices to offer nearly 50% more patient appointment slots than a control group.[6] Digital strategies can provide the needed efficiency gains.

COVID Inpatient Remote Patient Monitoring (IPRPM)

At first, adult inpatient COVID patients were discharged home with a specialty kit containing a Bluetooth-enabled tablet, blood pressure cuff, weight scale, and pulse oximeter device, allowing staff to virtually monitor and communicate with patients. As the initial supply of kits diminished, Virtua Health staff quickly discovered that the program was equally effective utilizing its video visit technology platform supported by a basic pulse oximeter. Those who needed additional support, such as those with chronic obstructive pulmonary disease and chronic heart failure, received the more robust kits.

Virtua Health nurses monitor patients remotely, supplemented by physician-led telehealth visits, as needed, to manage the patient population over the course of two weeks post-discharge. Results from IPRPM include a 32% relative reduction in hospital readmissions, a 97% rating from patients who felt more comfortable knowing a nurse was checking on them, and a 92% recommendation rate among patients for the program.

COVID Emergency Department Remote Patient Monitoring (EDRPM)

Patients with COVID diagnoses who are deemed appropriate for the EDRPM program are discharged home from the emergency department with a pulse oximeter and scheduled telehealth visits with a physician to monitor health status. During 2021, 1,100 patients were enrolled in the EDRPM program. Of those, patients experienced a 43% relative reduction in 14-day ED return visits.

Care After COVID, or Long Haulers Program

An integrated care program for patients experiencing continued COVID symptoms for more than 30 days, the long haulers program includes those with such symptoms as shortness of breath, fatigue, brain fog, and more.

Patients initially complete an online questionnaire and participate in a comprehensive initial evaluation via video consult. The number and timing of follow-up visits depend on patient progress, and the patient is released from the program when sustained improvements are seen.

Cardiothoracic Preoperative Introduction

Initiated as a way to improve network integrity, a patient whose cardiac catheterization shows abnormalities that warrant surgical evaluation is afforded the opportunity to speak with a cardiothoracic surgeon on-demand through a telehealth encounter. The virtual nature of the visit enables the patient to add family members or caretakers to participate in the encounter, which allows real-time feedback from the operating surgeon to assist in the patient's medical decision-making.

Although this encounter is not billable, the health system has the opportunity to maintain network integrity by 20–25 cases each quarter, which more than makes up the cost of the program. By conducting an additional visit at the time of service, patients immediately know the next steps in their care journey without the stress of uncertainty while waiting for another appointment.

Newer Programs Increase Patient Engagement

In early 2022, Virtua Health launched the Inpatient Warm Handoff to Outpatient Social Worker program, a state-mandated, quality-based payer reimbursement program for behavioral health and pregnant patients discharged from inpatient settings.

Patients leaving the hospital often find it difficult to follow up with outpatient social workers. Using a virtual visit, the social worker can have an initial conversation with a patient to identify and address any barriers to follow-up. The goal is higher adoption among these patient populations.

Virtua also recently launched a Primary Care Telehealth Practice and Neurosciences Care After Discharge program.

Success Stories and Lessons Learned

Virtua Health's digital transformation has been wildly successful among both patients and practitioners. The three COVID telehealth programs earned a

Net Promoter Score (NPS) of 75%, more than double the industry average for healthcare of 31%.

When deployed deliberately, digital tools can also have a positive impact on staff shortages. Because digital tools are intuitive, the DTO discovered that new clinicians could orient to telehealth technology using an on-demand 20-minute training video.

Virtua has extensively studied what other health systems are doing, not only to get ideas but to gauge their own digital efforts against competitors. When scoping a new project, DTO staff and stakeholders work in six- to eight-week increments, imagining what innovation can be achieved during that time period. They also conduct technology assessments in a similar time span.

Virtua leaders stress that digital solutions don't have to be perfect at the outset, a difficult concept for many healthcare leaders to grasp. As long as the solution is clinically sound and delivers value, even an 80% effective automation solution, for example, will benefit the vast majority of patients. That leaves 20% of patients for high-touch engagement.

Governance is an integral part of each project, ensuring not only project quality but also that the pace of change moves in step with the maturity of the transformation. A project that is quickly and widely accepted by staff in one area will need much less governance than one where adoption is slower and more difficult. And while viral adopters are integral to the success of any project, an organization can't move at a viral adopter pace with every project.

Prioritization and re-prioritization are key.

Each project represents a juggling act among the various stakeholders, and good governance provides the necessary cohesion to give each project the best chance for success.

AI Aids in Patient Selection for Hospital at Home Program

Rachini Ahmadi-Moosavi & Meera Udayakumar

Advanced analytics blended with operational and clinical workflow enhancements and technology assistance has become a transformational force in healthcare. This innovation story is just one example of how data and analytics are catalyzing changes to healthcare delivery.

UNC Health began its Enterprise Analytics and Data Sciences (EADS) department in 2016 to accelerate analytics adoption throughout the health system. Since that time, EADS has merged into the Information Services Division (ISD) to seamlessly join analytics foundations and infrastructure with analytics services and delivery. While system-wide analytics adoption lagged behind capabilities available to the health system for a few years, the pandemic became a catalyst for driving wide-spread reliance on insights to run, grow, and transform UNC Health. The pandemic provided clarity of focus—all leaders and layers of the organization addressing common challenges—which allowed analytics to shine as a beacon that aids in navigating uncomfortable and sometimes impossible situations.

Like the rest of the world, UNC Health faced serious clinical and operational challenges that led to rapid data-driven culture change. UNC Health quickly embraced the adoption of analytics to create many COVID-19 insights and initiatives, including the following:

- Forecasts and dashboards highlighting bed, staffing, and supply needs
- Multi-disciplinary work groups leveraging analytics to track supply chain shortages for PPE and contrast and to monitor vaccination delivery and impact
- Recovery analytics to aid in operational and financial performance

Virtual Care as a Disruptor to Traditional Care Delivery

In this post-pandemic world, U.S. hospitals are struggling to navigate financial and capacity pressures—fueled by higher labor costs, staff shortages and workforce burn out, lower reimbursement due to higher acuity, increased length-of-stay, supply chain disruptions and cost increases, lack of recovery of procedural and surgical cases, and bottlenecked transitions to post-acute care. It is becoming increasingly clear that for providers and consumers, the modes of healthcare delivery and demand have shifted to more convenient, less intensive, and in many cases virtual. In their 2021 book *Healthcare Digital Transformation: How Consumerism, Technology, and Pandemic are Accelerating the Future*, Edward W. Marx and Paddy Padmanabhan articulated trends that will disrupt traditional health system operations, specifically the growth in popularity and reliance on virtual care (Marx & Padmanabhan, 2021). Today, hospitals across the nation are attempting to shift medically appropriate patients to home-based care in lieu of formerly acute care services.

Award-Winning Hospital at Home Program

The UNC Health *hospital at home* program is branded Advanced Care at Home (ACH). ACH provides a combination of 24/7 virtual care and in-person acute care services to patients in the comfort of their own home. These services can include "infusions, diagnostic tests, nursing assessments and physician visits," and just like an inpatient stay, a multitude of care givers deliver pieces of the patient's care plan. UNC Health ACH deploys "a network of in-home service providers including paramedics, skilled nurses, lab services, mobile imaging, therapy, and more." Patients are considered "sick but stable" with "specific diagnoses such as heart failure, COPD, pneumonia," etc. (Phillips, 2022).

Since hospital at home programs are reliant on adequate reimbursement to fund the service, patient self-pay or insurance status has been a factor in selection for the program. ACH initially focused on early discharges for Medicare, Medicaid, and Uninsured patients who still required some degree of acute care services toward the end of their inpatient stay. The ACH program later expanded to direct admission from the emergency department for multiple other payors as the reimbursement situation improved. The program kicked off with two hospitals in the UNC Health system—UNC Medical Center in Chapel Hill, NC, and UNC Rex Hospital in Raleigh, NC—and quickly expanded to two additional sites in Hillsborough and Holly Springs, NC. Since the program provides in-person care, geography is a factor in who can quality for ACH. Even with ACH at only four hospital sites, "in its first year, the Advanced Care at Home program provided care to 456 patients ranging from age 37 to 103 across 46 zip codes in the state of North Carolina" (Phillips, 2022).

A UNC Health Newsroom article (Phillips, 2022) quoted Dr. Meera Udayakumar, Advanced Care at Home Medical Director, saying, "This care model allows patients to avoid certain risk factors inherent to traditional hospitalization, such as deconditioning and delirium or confusion." She added,

> It also allows caregivers to be more involved in a patient's hospital care, which is especially important if that individual is going to be supporting the patient after hospital discharge. Additionally, this program also provides the care team insight into a patient's home situation, which can have a significant impact on their health.

In August 2022, the North Carolina Hospital Association bestowed the 2022 Highsmith Award for Innovation to the UNC Health Advanced Care at Home program, marking a successful first year of operation.

Treating patients in the comfort and safety of their own homes through the ACH program has many benefits:

- Reduces the risk of nosocomial infections
- Lessens exposure to the multitude of infectious disease surges impacting local hospitals
- Frees up bed capacity within high-occupancy hospitals experiencing staffing shortages
- Reduces the stress that many patients face when being hospitalized, away from their loved ones and comforts of home
- Helps many patients maintain their independence and continue to heal at home

The article noted:

> *The Advanced Care at Home model enables patients to receive the right care in the right place at the right time. One year of success so far and the program has already shown improved satisfaction, comfort, and convenience for patients and caregivers.*

Acute care-at-home services will continue to grow as healthcare technology advances, patients become more comfortable with connected health devices and virtual care services, patient identification continues to improve, and programs can support more categories of patients. "The future impact of this approach can only pave the way for a better experience and a more personalized healthcare journey for patients" (Phillips, 2022).

Analytics Powers Patient Selection

While the hospital at home care delivery model is innovative by itself, the second part of this transformational program is the patient selection process. Assessing which patients are good candidates for hospital at home services can be labor intensive, requiring copious amounts of manual chart review to determine numerous clinical and social/environmental exclusions and geospatial proximity to hospital at home services. In recognition that speed to and accuracy of patient selection is a critical part of a successful program, Dr. Udayakumar and the UNC Health Advanced Care at Home (ACH) team partnered with the Data Science team in ISD EADS to co-create an AI model that produces a score indicating which patients are better candidates for the program. The first iteration of the model was released in September 2021 and assessed 61 variables. Via continuous improvements on the model, in April 2022, 13 additional variables were identified and included. While clinical predictive models do not remove the need for manual review and input from a clinician, it does allow for optimized identification of where clinicians should spend their time assessing patient fit for ACH selection. Essentially, the model acts as triage for evaluating patients' care plans, helping clinicians make more informed decisions.

The analytics model was released in a phased-agile approach, leveraging a product management discipline. The phase 1 scores were delivered to the ACH team outside of the Electronic Health Record (EHR). The second phase integrated the model's score directly into the EHR to enhance the clinician experience with selecting patients for ACH care. Additional phases of work are ongoing to further enhance the patient selection process and are defined in collaboration with the ACH clinical/operational team, EHR technical team, and EADS analytics teams.

The AI model has set the UNC Health ACH program apart from other hospitals that are pursuing similar endeavors. It has also gained its own recognition, and the model is under consideration for commercialization. The magic behind this successful implementation of an AI model has everything to do with collaborative partnerships that enhance clinical and operational programs and enact positive changes on the technology, in this case the EHR.

References

Marx, E. W., & Padmanabhan, P. (2021). *Healthcare digital transformation: Howe consumerism, technology, and pandemic are accelerating the future: HIMSS book series*. Boca Raton, FL: CRC Press.

Phillips, B. (2022, August 19). *Advanced Care at Home Program receives highsmith award at one-year milestone*. Retrieved from UNC Health and UNC School of Medicine Newsroom: https://news.unchealthcare.org/2022/08/advanced-care-at-home-program-receives-highsmith-award-at-one-year-milestone/

Innovation through Application Integration

Laurie Eccleston, Gigi La Course & John Goodgame

In healthcare, HL7v2 (Health Level Seven) is one of the most basic ways of exchanging clinical information or data between applications. It is a standard way of communicating messages between systems in a meaningful way that enables the improvement of clinical effectiveness and patient care.

An integration engine is one way to achieve interoperability. It removes the need for point-to-point interfaces, thus allowing the routing or transmission of one message from the sending system to multiple receiving systems. Integration specifications are usually provided by the vendor that document the data they are expecting in the message segments and the required or optional fields for exchanging data between the systems.

Integration requirements are often the business needs or problems that are trying to be solved through the transfer of patient data from one system to another via an HL7v2 message. Analyzing the business need, specific workflows, or expected outcomes is imperative for understanding the flow of data and integration design. Typical integration projects follow an iterate process such as shown in Figure 2.2.

While HL7v2 messages have been the dominant player, some new formats and protocols have started to emerge in the market. There has been a great deal of confusion around what each type of message and protocol is used for and when they should be used.

For example, HL7v2 messages have been used for years to keep systems in sync. An event occurs, such as an admission, discharge, order, or a result, and another system needs the information to keep up to date. The message is sent, and each system keeps track of the data, such as details pertaining to the visit, patient, insurance, diagnosis, observations, results, and the list goes on. While this works well for keeping systems in-sync, it's not very practical for mobile devices as the amount of data collected has grown exponentially over the last few years.

Figure 2.2   Typical integration projects follow an iterate process such as the one above.

The target audience of these transactions has also changed. The ancillary laboratory system that needs all admission messages in case an order is placed is very different than a person who may want to know the result of their COVID-19 test.

Although HL7v2 supports a variety of application integrations, how do you determine if HL7v2 is the appropriate method or solution for achieving the intended outcome? Are there other tools that can offer a seamless integration strategy, or is it dependent on the type of clinical data to be exchanged, software application, mobile devices, or another mode of delivery? Or can it be a combination of HL7v2, FHIR, FHIR API, or restful service?

Additionally, there are other technologies that are used for sharing data among software systems, including the following:

API: An API is a service that provides some limited set of functions when provided specific information, quite often through a URL. For example, the weather service API just needs a zip code, or a city and state, and it will return the forecast for that location.

FHIR API: These are well documented at https://hl7.org/fhir/, but it's simply accessing or updating data through a URL. For example, retrieving a patient might look something like this: http://usp5540jbgood:52781/csp/healthshare/fhirnamespace/fhir/r4/Patient?name=JohnBGood

This would return a patient resource for "John B. Good," which may be used to retrieve other FHIR resources associated with the patient ID.

Despite having these additional tools or methods, is it possible to manipulate and leverage traditional HL7v2 messaging to accomplish system integration, delivering results that drive seamless exchange of data for patient care? From an innovation perspective, we believe it's possible and offer the following example of how this has been achieved.

Monkeypox infection tracking is critical to any infection prevention department, much like COVID was during the pandemic. Unlike COVID-19, where test results are of significance, to fully flag and track monkeypox exposures and infections, both orders and results are needed from internal and external labs.

Monkeypox testing can be performed at many different laboratory locations. In this example for illustration purposes, we will call them Lab One, Lab Two, and Lab Three. Orders and results for Lab One and Lab Three are

sent through an integration engine. However, orders and results for Lab Two
go directly from an EHR via an API to a laboratory. Also, patient identifiers
for Lab One and Lab Two are different than that used for Lab Three and
by the vendor software used for infection tracking—clinical surveillance
software. Labs One and Two utilize an EHR's internal identifier, while Lab
Three and the Clinical Surveillance Software utilize an organization's medical record number (MRN).

- So how can you accomplish flagging of monkeypox exposures from
  both internal and external labs utilizing HL7 and an integration engine?
- How can you get orders for Lab Two to the integration engine in addition to the API so that the engine can route the order to the Clinical
  Surveillance Software?
- How can you include the MRN in the orders and results when sending
  these from Lab One and Lab Two to the Clinical Surveillance Software?
- How can you ensure that orders routed to Lab Three from the EHR
  were sent to the Clinical Surveillance Software in a timely manner for
  flagging?

First, we will tackle the orders innovation.

For Lab Two, there needed to be a way to somehow access the original
Lab Two monkeypox order and send it to the integration engine for routing.
It needed to be distinguishable enough so that the engine could identify it
and then send it on its way to the Clinical Surveillance Software.

To achieve this, build was completed in the EHR to use order transmittal
to duplicate the monkeypox order and send it to an outbound orders interface that connected to an integration engine while still sending the 'original'
order to the API connection to Lab Two.

The existing Lab Three interface for this purpose was used. Doing so
solved two issues: Since the order identifier for monkeypox is different for
Lab Two and Lab Three, the engine could use this value to identify and
route the order to the Clinical Surveillance Software and then prevent it from
going anywhere else. Also, the Lab Three interface specifications in the EHR
uses the MRN, which can be ingested by the Clinical Surveillance Software.
Therefore, the Lab Two order would also contain this identifier.

For Lab One, additional routing in the integration engine was required
to send the order to both Lab One and the Clinical Surveillance Software.
This was a straightforward process. However, since the Clinical Surveillance
Software is expecting to receive the MRN, not the EHR's internal ID, we

had to somehow get this in the message. This was achieved by changing the specification of the outgoing orders interface in the EHR to include both identifiers. Code was then added to the engine to strip the unneeded identifier for each receiving application and only send what was expected by both.

To make this process even more seamless, the monkeypox test identifier in the messages was edited in the engine to reflect what the Clinical Surveillance Software was already receiving from Lab Three. Doing this meant no additional build in the Clinical Surveillance Software.

For Lab Three, we needed to ensure that the department received timely notifications of monkeypox orders. To achieve this, the engine added code to route the EHR order to both Lab Three and the Clinical Surveillance Software based on the order identifier in the message. This allowed the Clinical Surveillance Software to be notified at time of order, rather than the receipt of the specimen in the lab.

Second Monkeypox Results

For results from Lab One and Lab Two, we reused two existing 'rebound' interfaces in the EHR that were created to allow COVID results from Lab One and Lab Two to be sent via the engine to the Clinical Surveillance Software. These interfaces receive rebounded incoming COVID results from the two labs and send them outbound to the engine where they are routed to the Clinical Surveillance Software. Because the interface specifications already used the MRN, this only required adding the monkeypox identifiers to a rule that identified what tests to rebound from the inbound results interfaces.

Like it did for the orders from Lab One and Lab Two, the engine edited the result identifier to mimic what the Clinical Surveillance Software was already receiving from Lab Three.

This innovative achievement demonstrated a way of leveraging interfacing/integration technology that yielded an automatic flagging versus manual intervention in the Clinical Surveillance Software. Ultimately, this was faster and safer by eliminating manual work and marking patients with orders for appropriate isolation, thus reducing exposures and exposure workups.

Newer technologies are more geared to light weight service models, or a "pulling", more than "pushing", for every data event that occurs. In the case of FHIR, it's really about a storage and retrieval of FHIR resources. A slice of data that's related is considered a FHIR resource, a patient an encounter, and so on. This is easier for applications to ask for a specific document or a specific encounter rather than an entire medical record. APIs and restful services are just a mechanism that exposes methods that retrieve or post information to these FHIR resources. FHIR and APIs are designed for the light traffic a mobile application really should be using.

Despite these newer methods of interoperability or connectivity for data sharing, HL7v2 isn't going anywhere; it's the best model for pushing information between systems to keep them in sync. Larger enterprise systems such as an EMR or Datalake are much more capable of providing real insight into patient care or providing AI/ML analysis to improve patient care, but as we integrate the patient into the real-time communication and insights, these larger systems are playing a crucial role that allows the analytics, machine learning, or population health algorithms to run on the enterprise and provide the patient or mobile user with notifications in real time as a service.

In summary, what we are starting to accomplish with the interoperability systems is recognizing the value of the different solutions and the usefulness of each technology for a given situation. In short, the HL7v2 event transaction processing is extremely efficient for keeping core systems in-sync and capable of providing significant innovation.

The newer technologies FHIR and FHIR APIs are well-designed to integrate patients and expose services to their mobile devices and provide a lightweight communication for facilitating the integration of the patient into participating in their medical care. Looking forward to the next few years, it is likely that the medical applications on our phones will let us know we are overdue for a lab test as this is happening now with several facilities.

Maybe we can even hear our phone saying, "There is a 3:15 opening at the lab located across the street from the location of your 2:00 P.M. meeting. You have a 45-minute opening before your next meeting. Would you like me to schedule that laboratory visit for you?"

* * *

Innovation through Technology-Powered Practice Management

Shawntea Gordon

Information technology (IT) and artificial intelligence (AI) are entering the healthcare industry in full force, causing apprehension, excitement, and opportunity to abound. The idea of treatment decisions driven by AI using predictive analytics and an enterprise data aggregation system sounds both evidence-based and frightening. There is definitely something to be said about the human element. In 1911, Elbert Hubbard once said, "One machine can do the work of 50 ordinary men. No machine can do the work of one extraordinary man." More than 100 years later, this statement still resonates because we recognize individuals are necessary to the element of care delivery, but the same goes for practice management.

IT is developing so quickly, in fact, that most times it is outpacing the ability to regulate it. This is true across various industries, especially healthcare, with the greatest difference in implementation being that most of us in healthcare have yet to optimize the use of technology. Observing activities, people, processes, and technology is important, but it is the data that tells the full story. As Daniel Keys Moran so aptly stated, "You can have data without information, but you cannot have information without data." The best way to get to the data? People, processes, and technology, a cycle that can either taste bitter or sweet depending upon your level of engagement and intention.

We are entering a world in which we need technology to *support* us. Alper, Elliott, et al. performed a study in 2004 that found physicians need to read 29 hours a day to remain current in their field.[1] That was almost 15 years ago. Those numbers have undoubtedly increased as science and research continue to expand. Obviously, this is humanly impossible to achieve without support. This is an area in which we need to lean on technological capabilities, and many around the industry have done just that.

At a basic level, we have things like 'clinical decision support software' and 'preventive service reminders.' On a more sophisticated level, we have examples like organizations using IBM Watson to provide a second opinion or AI-mining databases to identify diabetic patients and predict which ones are most likely to have kidney damage. This is the more comforting side of technology, using it as a supporting tool. Less comforting would be for technologies to take it a step further by physically speaking to the patient and guiding them through their treatment.

Undoubtedly, patients in distress would feel more comforted by a nurse than by a robotic voice. Human interaction matters. Inflection and tone of voice, body language, and discrepancies between the two are things that AI struggles to emulate.

There are two questions that make frequent appearances at brainstorming sessions in healthcare technology: "Could we?" and "should we?" The ability to run algorithms and projections in less than a minute makes technology the optimal resource for answering if something *can* be done. The morality and empathy as to whether something *should* be done will probably always require a human being.

The goal is to bridge the gap between providers, technology, and patients. Guided by physicians and empowered by technology, staff provide the human element that creates the atmosphere for patient engagement and strategic planning. The technology platform should support care delivery, financial stability, and operations automation: Technology-powered management.

With technology-powered management, we can create strategic, structured, and repeatable services. For example, we can improve alignment between provider treatment goals and patient performance and compliance. Cisco Internet Business Solutions released a report noting that "the average person is projected to have between six and seven connected devices by 2020."[2]

Meanwhile, recent *JAMA* articles have performed studies into stand-alone mobile health interventions. A recent publication[3] focused on the potential impact that blood pressure wearables could make for hypertensive patients. The results were insignificant: "Use of a smartphone app resulted in a small improvement in self-reported medication adherence but did not affect blood pressure; the benefit of this and other stand-alone mobile health (mHealth) interventions on clinical outcomes remains to be established."

To drive efficiency, companies must focus on incorporating innovative design, technology-powered care processes into their foundation. Engagement and all-hands brainstorming sessions guided by advanced team leaders with cross-functional SME experience alongside full executive support make a significant impact.

All-Hands Meetings

There is a movement in healthcare, originally meant for patients, referred to as "nothing about me without me." Meaning if a change will impact me, I should have a seat at the table. This thought process is present in business as well. If a process change, new technology, or revised automation is going to affect a specific department, then they need to be a party to the discussions.

Take it one-step further with "all-hands" style meetings, "nothing about us, without us." All-hands meetings aren't new, "the name 'all-hands meeting' comes from the phrase 'all hands-on deck'—a signal that requires all ship crew members to go on deck" (Slido). Used in professional settings ever since, this also picked up speed when Silicon Valley began using all-hands as a go-to communication method for their rapidly growing businesses. The benefit of all-hands meetings is multi-faceted, but the driving force is increased and consistent communication. Except for HR meetings and board meetings, consider having an open-door policy for all meetings. Meaning, if I see a meeting of interest or concern to my department, I can invite myself to the meeting. For those considering implementing, I highly recommend it, provided there are ground rules. I can't take over the meeting, I can't attend simply to shoot down the suggestions of others, etc. But I can attend and participate. I can identify potential program hurdles, offer alternative solutions, and volunteer to support within my field of expertise. This is a significant driver for innovation. Let's look at what drives innovation: Personal experience, choice, and the rewarding feeling that you contributed to something in which you desired to participate.

If you've heard Dan Pink's talk about motivating factors for employees, then you can understand this is highly motivating to employees. Basically, he identifies that monetary rewards for thought-based achievements only incentivize at a small rate. The greater motivation for employees is the opportunity to do something that they believe to be personally rewarding. This falls in line with the top of Maslow's hierarchy of needs, which also define motivating factors for employees.

These meetings give every employee, from the front desk up, the opportunity to participate in company-wide change. Getting views and input from all levels mitigates issues before they occur. For example, in 2015, France spent the equivalent of $15 billion updating their railcars. Unfortunately, it wasn't until after the railcars were created that they realized they wouldn't fit at 1,300 of their existing railway stations. This cost France an additional $55 million and begs the question: How many people could've identified the issue ahead of time had they simply been included in the conversation?

Cross-Functional SME Leaders

When we first enter meetings, the thought process is to leave your business card at the door. Your title alone does not make anyone else's suggestions immaterial. The goal is to suggest opportunities, ideas, and solutions. So, it's important that discouraging language and demeaning behavior is strictly prohibited. Once that's established, a free-flowing brainstorming session can begin. The more variety of the views participating, the better. However, these 'all hands' meetings could quickly get 'out of hand' if we didn't have the appropriate leadership in place.

Each meeting has SME leaders that help to define the problem or refocus meeting attendees as needed. They provide the structure of documenting the thoughts, steering conversation in optimal directions, and making sure voices are heard. These are often the people delegated to perform the project management and action planning tasks. Leadership experts know how to create meaningful thought exchanges and are able to observe who has the most to contribute to the project for their planning going forward.

Full Executive Support

The ability to innovate, positively disrupt, and implement systemic change is not possible without the full support of the executive team. Without the approval to spend time and resources on an innovation project, any tasks toward it will be viewed as "off mission." One of the primary roles of the executive team is to ensure that organizational goals and operational tasks support programs that are accretive to the mission of the organization. So, it's important for employees seeking executive support to keep the corporate mission in mind.

Experienced innovators know that executive sign-on is mandatory. Especially because the executive team usually has a higher-level view of the entire operation and can visualize applications of solutions across other departments. For example, let's say I notice my billing staff makes up 10% of the organization and is experiencing reduced productivity. I have an all-hands meeting, and with our SME leaders, we identify a scrum-style software application that could provide a productivity increase of 15% in my department. I then gain executive support to prioritize with the IS department because this task management program could also support the clinical staff, human resources, IS, and administration departments. This changes the project impact from 15% productivity increase in 10% of the organization to 15% productivity increase across 100% of the organization.

When innovation and continuous improvement are part of the mission, make sure to allocate resources like staff support, funding, and project managers toward project achievement. It prompts questions like the following: "What do you estimate this would cost to achieve?" "How can we allocate funds?" "How can existing or new staff support this?" When these are not part of the mission, then those resources instead become barriers to innovation and improvement, prompting conversations that include statements like "the way we do it is the way we always have," or, the worst I've ever heard, "there is no line item for 'Innovation' in our budget."

Innovation has to be systemic to an organization. This applies to every industry but especially healthcare. Especially right now. Our industry is decades behind in utilizing the integration technologies that other industries rely on daily. Using supporting technologies could enhance access to care, price transparency, and coordination of care.

If Retail Worked Like Healthcare

If retail worked the way healthcare does, Amazon wouldn't exist. Imagine that to buy a new outfit, you first had to sign an agreement to have access to your previous size history. Even though it is your own history, you need permission to access it. You sign permission and complete a few required forms, and a few weeks later, you have information on what once fit you. So then, you contact a store that has excellent reviews, and they pencil you in for three weeks from now. You arrive at the store and wait for 30 minutes while filling out forms that are identical to the ones you filled out online when you requested your size history. Finally, an attendant takes you back, looks you up and down, mumbles 'okay, thanks,' and leaves. Then a stylist walks in, looks at her computer, says, "wear blue" and leaves. You don't have a clue what that means, you're not sure what anyone typed into the computer. Six weeks have passed, and you still don't have a new outfit! This isn't the Amazon of today for a reason. People want some element of technology, they want transactions to happen quickly, and they want history to pull up at lightning speed.

People also want people. Amazon had 4% of overall retail expenditures[6] across the United States in 2017, but they also had more than 500,000[7] employees dedicated to service delivery, problem resolution, and predicting needs. For healthcare 2.0 to provide shorter wait times, increased visibility, price transparency, predictive analytics, and increased patient touch points, we need to support our clinical providers with technology-powered care services. We need to get innovative technological tools in the hands of our clinical staff to enhance how they carry out orders from providers.

Ways to Do Things Differently

Look at other industries; they have been mastering technology implementation and strategies far longer than we have. Many of their solutions are easy to implement, scalable, and yet radical to this industry.

It is normal to be apprehensive about these seemingly mythical solutions from other industries, but keep in mind, these aren't new ideas. Some of these technologies have been around for decades.

> *The gas company can read my meter from 300 feet away and*
> *issue me a bill within seconds that identifies if my usage is higher*
> *than my neighbors. Why can't my watch report my pulse rates to*
> *a dedicated system at all times and immediately notify someone if*
> *I am suddenly in A-fib?*
> *   Why can't my practice management system let me know that*
> *I've received underpayments from eight claims and send me a*
> *summary of the information needed to investigate further?*

Actually, with technology, these things can happen; these things are possible. When we open our minds for innovation, get executive support, implement new technologies and focus all of this on a motivation to improve the patient experience, brilliant things can happen.

Today, in less than a minute, I can look at an entire provider roster, identify patients eligible for chronic care management, stratify those with the highest risk of exacerbation, identify likelihood of upcoming hospital admission, sort based on barrier to care, pull education with patient history, and assign them to a high-risk population for preventive engagement designed specifically for them. In less than a minute.

In seconds, I can pull a dashboard of denial reasons and implications by department alongside trends and future projections. I can export work lists for A/R or denial management to my staff and track their progress in real-time. I can even run an analysis of my entire patient database and get a report on demographic inaccuracies or duplicate records.

Apply by Focus Areas

We can review technology optimization in whole or by area of focus. Let's take RCM, for example: A quick internet search easily confirms that there are hundreds of revenue cycle management technologies available. An exciting and daunting amount of options for review that each contain their own host of widgets, colorful screens, and automations. Another internet search provides a hundred articles on the best ways of selecting technologies with pros and cons that range in bias levels from slight to moderate to extreme.

The most critical component is neither the quantity of offerings nor the opinions of others. In fact, the most important factors for success are understanding your own needs and understanding the technology chosen. Merely using a practice management system or clearinghouse in itself is not innovative. Identifying key performance indicators (KPIs) and documenting acceptable use and entry of data are standards that should occur in all facilities. Technology is present and ready for use, and how we use it is where we have the opportunity for innovation.

Every piece of data that we obtain, enter, or ingest is a potential puzzle piece in our larger RCM picture. To capitalize on the creativity of our workforce, we must provide them with the technological tools, training, and automation necessary to achieve success. In RCM, that means comprehensive evaluation of every component in the revenue cycle.

With a focus on RCM, I may look at denial trends by department or by category. I may automate the alerting of underpayments or over-coding trends. I might develop strategies to more quickly identify and work denied claims through automated work list creation for my staff.

Without technology, this wouldn't be possible. Without people, this wouldn't be achievable. The goal is to continuously improve this method of technology-powered management. Our patients need it, our providers need it, and healthcare deserves a technology-powered service delivery revolution.

References

1. Alper BS, Hand JA, Elliott SG, et al. How much effort is needed to keep up with the literature relevant for primary care? *Journal of the Medical Library Association*. 2004; 92(4): 429–437.
2. mHealthIntelligence.com. Interventions need care team support to affect outcomes. 2021; 39.
3. French red faces over trains that are 'too wide'. www.bbc.com/news/world-europe-27497727
4. Amazon grabbed 4 percent of all US retail sales in 2017, new study says. www.cnbc.com/2018/01/03/amazon-grabbed-4-percent-of-all-us-retail-sales-in-2017-new-study.html
5. Amazon now has more than 500,000 employees. http://money.cnn.com/2017/10/26/technology/business/amazon-earnings/index.html

* * *

How Conversational AI is Bridging the Communications Gap and Automating Contact Center Operations

Dan Howard

I have always been a big movie aficionado, especially those in which the technology is always years ahead of its time. Some of my favorites are *Star Wars, War Games, TRON, The Matrix, Terminator, Star Trek,* and *Minority Report,* to list just a few. These films share a common thread where the use of artificial intelligence and advanced robotics was commonplace and had the ability to complete complex tasks and cognitive interactions beyond the level of any human effort. Fast-forward to present day, and we are all experiencing how modern-day life is beginning to mimic the science fiction seen in movies and television. Most of us carry around personal, digital assistants that can be easily summoned, by simply speaking out loud, that fulfill a variety of tasks or provide information based on our queries. From social networks to mobile messaging apps, we are getting more and more done using these specialized applications that are harnessing the power of artificial intelligence.

The science of artificial intelligence (AI) goes back a few years to its inception in the mid-1950s. It was originally defined as the ability for machines to imitate or display human cognitive traits or skills involving the ability to solve problems, complete complex tasks, or learn something new. Most modern definitions have restated this definition and now loosely define AI as the ability for a non-human system to rationalize and take actions that have the best chance of achieving a specific objective. Retail and e-commerce, banking and financial services, logistics and transportation, entertainment and gaming, and manufacturing are some of the industry verticals that have advanced their competitiveness by infusing AI into their operations. We see this increasingly in areas where customer-facing interactions are key to enhancing the consumer experience. For example, many of the digital chatbots available on commercial websites are AI-powered and designed to provide instant answers to a range of common customer questions and intuitive product suggestions. Healthcare has made significant advances with AI in several areas, but most are not truly customer facing. These include pattern recognition in data mining for more accurate diagnosis and treatment of medical conditions, robotic surgery, drug research and discovery, RPA for revenue cycle, medication management, and others. Healthcare continues to lag behind some of these other industry verticals, but the good news is that there continues to be a strong investment by Google, IBM, Apple, and Microsoft and even more venture capital funding of the many new tech startups that have a focus on digital, omni-channel consumer interactions. We live in a digital world, and customers are preferring to engage in digital channels for self-service tools and when looking for answers or resolution to simple issues—but they still also want that human connection.

San Ysidro Health is a large FQHC health system of clinics and urgent cares located in San Diego County. Our services footprint spans across the county's 4,300 square miles and covers a population base of 3.2 million residents. We pride ourselves as being the safety net for the safety net providers (county hospitals, etc.) in the region where we provide high-quality, compassionate care to the underserved, undocumented, and marginalized communities. Prior to COVID-19, the discussion around digitalizing the patient experience was happening, but the urgency was missing. As the pandemic intensified, we realized that we needed a digital transformation, and what was once considered a nice to have now became an operational imperative. So, telehealth rapidly expanded and remote work venues were established as we pivoted quickly to meet the needs of our patients and providers.

An area in which we were continually challenged was our patient contact center. Patient calls during the last two years have grown exponentially and far outstripped the capacity of our call center agents. Patients were waiting on hold for long periods of time for all manner of requests: Appointment adds or changes, Rx refills, general information, lab results, release of information, directional wayfinding, etc. The continuing challenge of keeping and retaining contact center staff only exacerbated our call abandonment rates, decreased patient satisfaction scores, and delayed specialist referrals. The following scenario is how we addressed some of these problems, not by simply implementing a technical layer on top but rather interweaving the technology, conversational AI, and the live contact center agents into our desired digital communication platform.

We knew that conversational AI—think of advanced versions of Siri or Alexa—had to be part of our solution. We had three initial KPIs that would validate the effectiveness of the solution:

■ Reduce our call abandonment rate from an average of 19% to < 5%.
■ Reduce the hold time from upwards of 40 minutes to < 30 seconds.
■ Increase patient satisfaction scores in relation to contact center communication.

The platform would need to automate conversational experiences at scale via chat and voice-based virtual assistants working in concert with live human agents. It also needed to support our five core languages of English, Spanish, Vietnamese, Tagalog, and Arabic. The conversational AI market is not flooded with competitors yet as it is still a relatively new technology, but through our selection process, one solution began to stand out as best suited to meet not only our current use case needs but also many future ones as well.

Kore.ai was selected as our partner on our first, real, AI journey. Kore.ai is not the newest entrant into the conversational AI space, but they are listed as a top leader according to the Gartner Magic Quadrant for Enterprise Conversational AI Platforms 2022.[1] They really seemed to understand our need for an omni-channel, conversational, and experience-driven platform. We needed conversational AI-based automation and augmentation to support our live contact center agents.

One of the more refreshing things about embarking on a complex project such as this was that the core technology supporting the solution was already baked into the cake so to speak. This was a no-code AI conversational solution that allowed us to spend the lions' share of our time in the design phase. The unified platform supports the following key areas.

Conversation Builders

We had quite a few use cases and prioritized our list to those most important to operations and patient engagement goals. The platform allows for rapid and integrated conversation design with local IT and contact center staff collaborating around building dialog tasks, and robust virtual assistants to address our existing and new use cases in very little time. After initial training we were soon designing, building, testing, and deploying customized virtual assistants and digital tools that are delivering a variety of experiences for our patients today.

Multi-Engine NLP

The heart of the technology, where all the magic happens, is in the multi-engine natural language processing (NLP) core. This allows for very accurate outcomes that can intelligently handle the complex human conversations with the AI. It also machine learns as it goes, gaining meaning and knowledge from each patient and contact center agent. The innate ability for an AI voice agent to automatically recognize a caller's spoken language and begin a dialog with them in their native dialect was a game changer for us. But not only language, but intent recognition and caller sentiment. More on this later.

Omnichannel Experiences on Any Tech Stack

We had a need to deploy across multiple channels—mobile, web, digital app, and voice, and across other technology stacks. Our biggest integration was with our existing telecom solution contact center desktop applications and electronic medical records systems. API calls support these interactions seamlessly, and the fluidity of data between them is amazing. Switching channels, say from mobile to voice or web chat bot to live contact center agent, is done efficiently and in real-time.

Actionable Insight Reporting

I've always heard if you don't measure your results, regardless of your outcomes, you are missing out big time. Measuring and managing for results enables an organization to use the results information to understand what is working and what is not, and adjust accordingly. Our AI solution not only provides measurements in real-time but also allows for performance improvement adjustments in real-time. Dashboards provide metrics on bot, virtual assistant, and live agent performance.

So, what does this all look like in action in the real world? It is a game changer, at least for us and the patient communities we serve. I'll describe a very common interaction between a patient, our virtual assistant, and a live agent and how the AI orchestrates the entire experience end-to-end. Good experiences equate to good outcomes.

- Jorge is a 65-year-old Hispanic male with a history of hypertension and diabetes. He recently visited his primary care physician and was scheduled for a follow-up referral visit with a specialist to help better manage his blood sugar levels. Jorge has forgotten his upcoming appointment date and time and is also in need of a prescription refill on his hypertension medication.
- Jorge decides to call the San Ysidro Health main contact center line. Our existing interactive voice response platform answers his call, provides our normal canned response to call 911 or go to the nearest emergency room if you are experiencing life-threatening symptoms, and

then inquires whether he is calling for an appointment or something else. Jorge chooses appointment by pressing 1 on his touch tone phone, and he is immediately greeted with the following:

- "Hi, my name is Nancy, I am your virtual healthcare assistant. How can I help you today?"
- Jorge responds in Spanish, "Necesito una recarga de medicamentos y necesito verificar mi proxima cita." In English this translates to "I need a medication refill and I need to check my next appointment."
- Nancy immediately detects Jorge's spoken language and begins conversing in Spanish. The A.I will first validate Jorge's identity, through DOB, phone number and MRN number. This validation against the EMR record is nearly instantaneous—less than two seconds—and Nancy responds with "Veo que tiene 1 recarga restante para Zestril 20mg dos veces al día. ¿Debo procesar esta recarga?" (I see you have one refill remaining for Zestril 20 mg twice daily. Should I process this refill?)
- Jorge confirms the refill, and Nancy sends an automated message to the pharmacy system where Jorge's prescription is placed in the queue for refill. The system then sends Jorge a text message that his prescription refill is in process and will alert him when ready.
- Nancy then tells Jorge that his next upcoming appointment is for the next day at 8:30 A.M.
- Jorge is immediately upset because he has a conflict and tells Nancy that he needs to reschedule. Nancy looks for nearby dates and times and offers these appointment options to Jorge, but he does not confirm. Nancy inquires again, but Jorge still does not respond.
- The AI behind Nancy has picked up some tension in Jorge's voice and now his lack of responsiveness. Nancy asks if Jorge would like to be transferred to a live agent for assistance, and Jorge confirms yes.

This is where real-time insights are so valuable. Immediately after Jorge's confirmation to have his call transferred, a set of complex actions fire off. The AI solution hands off his digital voice channel to our contact center platform. Because the AI detected that Jorge's sentiment is possibly upset, his call is prioritized and immediately routed to a live agent. The agent's job of fielding this call is enhanced ten-fold because of the companion AI widget that sits on the agent desktop. As the call is transferred, the entire transaction is presented to the agent. They know that Jorge is the caller, why he is calling, that he already had the virtual assistant process a refill prescription, and that he may be upset about scheduling an appointment. The entire conversation is presented to the agent so they can begin assisting the patient without having to re-verify identity or ask other questions. Another advantage is that the AI desktop widget continues to listen in on the live conversation and bring up pertinent info from the EHR or other systems, saving time and freeing the agent from doing these lookups manually. If Jorge asks the agent where the address of his upcoming appointment is, the AI immediately brings up the address in the desktop widget and the agent can push that notification to the patient's mobile phone or e-mail address with an embedded map and directions.

This is but one use case of about 15 that we are using today. Many virtual sessions are completed end-to-end with no human interaction, while others, either based on complexity or patient preference, get handled in part with a live agent.

Measuring Outcomes

We are encouraged by our results. After just 90 days post activation, our KPIs speak for themselves:

- Call abandonment rates dropped to around 8%.
- Hold times to speak with a live agent reduced to an average of 20 seconds.
- Live agent productivity and availability to answer calls has increased.
- Extra costs were avoided by not having to hire additional live agents (equivalent to 11 FTEs).
- Patient experience scores will hopefully increase (still waiting on data and the passage of a bit more time).

We still have a long way to go, but we are innovating on the fly and adding new use cases each month. We see other areas of the business that could benefit across IT service desk and HR operations in support of our internal customers.

I am still enamored by those science fiction films from a few years ago and marvel at the technology presented—but glad to see that in some areas, reality and the technically out-of-reach are aligning to bring us solutions we need to improve the lives of our patients.

Reference

1. www.cxtoday.com/data-analytics/gartner-magic-quadrant-for-enterprise-conversational-ai-platforms-2022/

* * *

A Two-Way Clinical/IT Bridge

Donna Walker

I am a registered nurse (RN) clinical operations analyst and have successfully served on projects as a clinical bridge with IT, obtaining meaningful results and deliverables on special projects. I am sharing some examples in hope that other organizations will replicate them and harvest the rewards.

I was promoted from senior nurse auditor to a management position at a hospital where I previously worked. I was tasked to work with IT to grant access to a group of externally based clinical reviewers who were taking over the work from my earlier role. We had to create unique temporary accesses to certain EMR modules and applications at our hospital, being HIPAA secure, regulation/policy compliant, and having the functionality for the users to perform the work. I was able to effectively communicate the needs between the IT analyst and the clinical review staff. I explained to IT what the reviewers needed in functionality and then explained to the reviewers what IT was doing in the build process behind the scenes. Problems were conveyed to me from both the IT and clinical review sides to assist in developing solutions to resolve them. The IT analyst and I worked together to create training materials for the reviewers. We utilized conference calls, WebEx training sessions, screenshots, and email to communicate and evaluate our progress. This required open communication, translation of technical/clinical terminology from both sides, as well as a lot of tweaking by the IT analyst to finalize the product. The result was a HIPAA-compliant electronic system that limited access with temporarily created internal email addresses for the reviewers, and selective unique views created by IT with proprietary templates for review and response letter generation.

I conducted research and educated myself on some basic IT fundamentals to adequately perform in this bridge role. This augmented my communication ability with the technical side to in turn work with the clinical reviewers and bridge the information back and forth.

This example led me to continue working as a consultant on security risk assessments for HIPAA compliance for a large group of hospitals and clinics. I served on a diverse team consisting of IT, IT security, clinical, and compliance professionals. We started the process by conducting a situational analysis with the C-suite and then reviewing policies and procedures that the health system currently had in place. We compared the findings against the HIPAA Privacy Rule, 45 CFR 160–164. I again served as a clinical bridge between clinical and IT employees. Hospital and clinic site visits were arranged where walk-about assessments were conducted, accompanied by local IT management. Observations of the physical environment, including work areas, workflows, and end-user activities, were also noted. Management employees were interviewed facility-wide, inquiring as to what they did to secure protected health information (PHI) and ensure HIPAA compliance, utilizing open-ended questions. The findings were compiled, written up, and presented to the corporate organization with suggestions for process improvement. Follow-up assessments were then arranged with internal management and staff.

Getting buy-in from both sides was found to be a bit challenging, as clinical and IT speak very different languages. Clinicians are a blend of art and science, thinking in creative ways. Medicine is not absolute as no two patients with diabetes or heart disease are exactly alike, for example. Information technology is more mathematical and linearly thinking, with strict adherence to algorithms. This is necessary in computer science and data analytics. It was found that there was significant opportunity for improvement in collaboration between the two disciplines. I took a helpful approach, attempting to answer the question of, "What's in it for me?" for both clinical and IT. Clinical staff are championed with the task of maintaining, improving, and optimizing the patient's health. I used the analogy of the health of the patient's medical record (EMR). An electronic medical record number is more valuable than a credit card number monetarily. Many clinical employees I interviewed were not familiar with this but were interested and appreciated the knowledge. EMR safety and health also applies to the employee's own medical record. It was observed that employees grasped the importance of PHI and keeping it secure more readily when they could assimilate an example that also applied to them as individuals. This definitely added value and assisted in effective learning.

Information technology desires accuracy of data and data transmission, with high concern for data integrity. People can help IT with this process and goal. I encouraged IT to collaborate more with the clinical members of the hospital team, advising them that clinical staff would be receptive. I encouraged them to "think different" and not get lost in translation with the end users. Simplicity in communication goes a long way as sometime less is more. Engage the people, for they are the customers. People are not the problem— they are the solution! Clinicians are helpful by nature, and most of them have answered a calling to be a healthcare professional with a desire to serve.

They want to help and will help to fulfil the promise of information technology in healthcare if included collaboratively to care for the health of the EMRs. Healthcare providers want to be part of the cure, not part of the disease!

I have found that some of the best ideas are generated by very diverse teams where the members have different talents and skills. The members do not all think alike, but, if they can find a common ground to work together as a team, tremendously innovative outcomes can be achieved. That common ground is the desire for process improvement and to find a better way of achieving winning results. I enjoy being a clinical bridge, and though it is quite a challenge at times, I feel it is my calling to serve in this much-needed capacity. Utilizing a clinical bridge breaks down communication barriers and improves understanding, not only between clinicians and information technology but between other departments as well. The clinical bridge can also assist in C-suite communication by adding more meaning to financials. What is the clinical benefit of the upgrade in software? Will this save in time and money and improve patient care for our customers? Why are the training costs essential in the implementation process? Are there new, more efficient ways for the clinicians to effectively document, freeing up more of their time for the patient? What does a security breach cost? For example, what is the financial impact of forgetting to wipe medical records from a rented photocopier prior to its return? It could be very costly because HIPAA violations are calculated for each and every record of personal health information that is compromised.

My suggestion for continued innovation is for healthcare organizations to utilize more two-way bridges between clinical and information technology. This is simplistic, but powerful, and delivers positive results. Maintain these bridges, travel them routinely, and appreciate the culture of both sides as they unite and become strong allies to fulfil the promise of information technology in healthcare.

* * *

Patient Safety Innovating in a Public Academic Hospital in Spain

Juan Luis Cruz & José Luis Bueno

One of the worst things that you can probably experience as a hematologist responsible for the blood bank in a hospital is receiving a call that a hemolytic transfusion reaction has occurred due to the incorrect identification of a patient. It happened in 2012, and fortunately the patient survived, but the whole team pledged that it had to be the last error they would ever witness.

Puerta de Hierro University Hospital is an academic public hospital located in the northwest of the Madrid region in Spain with more than 600 beds and more than 3,000 staff. The hospital is a flagship in Madrid's health system, which is an integrated delivery and financing system that serves more than 6.5 million people with 430 primary care centers and 35 hospitals with more than 15,000 beds. Despite being one of the best hospitals in Spain, we make one of those life-threatening errors every 11,000 transfusions, or about one or two per year. About one in ten of those errors results in the death of the patient.

It all started with a patient safety problem and the commitment to solve it; however, that was not the only problem with the transfusion process at the time. Hospital leadership created the Hemotherapy Commission in 2012 in order to oversee the complete process review involving all stakeholders, including prescriptors from the inpatient, surgical, and ambulatory settings, including nurses and hematologists. There were some key aspects that needed to be addressed, such as the following:

1. Optimizing blood component use. Blood components are a scarce resource that rely on donations. Public policies usually encourage donation, but few of them are in place to address the need for efficiency in consumption. A good blood donation policy in a country is to get around eight donations per 1,000 citizens.
2. Improving the detection and analysis of adverse reactions (hemovigilance). Nurses or MDs who carry out transfusions are responsible for identifying and reporting adverse reactions in passive hemovigilance programs. However, some adverse reactions are not recorded, especially those occurring after finishing the transfusion. Despite immediate severe reactions that were usually reported, it seemed many others remained hidden. Regarding late severe reactions, like transfusion-related acute lung injury (TRALI) and transfusion-associated circulatory overload (TACO), the hospital reported virtually zero incidents in the period 2010–2012, far different from country-level statistics that set a mean of about one per 1,200 transfusions.
3. Excessive workload for all actors involved due to an incomplete IT-supported process, with several systems involved, and reliance on paper and phone. As a non-exhaustive example:
   a. Doctors had to register an electronic prescription with many required fields that made the process painful and slow, and then print it and

give the paper to the nurse (or just leave the printing work for the nurses).

b. Following the prescription, nurses had to conduct a blood extraction (two extractions in case the patient was a new patient for the blood bank, checking to make sure the patient blood type is the same as the one previously extracted so to avoid a mislabeling in the blood tube collected from the patient), regardless of the availability of a still valid sample for that patient in the blood bank. They had to call the blood bank to ask (slower) or simply do the possibly unnecessary extraction (faster). Also, nurses labeled the tubes and had to call the blood bank several times to inquire where the process was. Once they received the blood components and did the visual checks needed (sometimes in duplicate), they proceeded with the transfusion. When the transfusion ended, they had to write down (by hand) a final report provided with the blood component by the blood bank (called "DCT"), including times, identity, numbers of the components transfused, and adverse reactions observed. The empty bags, along with the handwritten document, had to be sent back to the blood bank by means of a hospital porter. This procedure is a mandatory regulation for blood transfusion in most developed countries, including European countries and the United States.

c. Things were not better at the blood bank. They had to manually register every new prescription using the printed one received with the blood sample, re-label each tube with their own internal code, and register lab tests to be completed in a different system. After the compatibility tests were performed and the blood components were selected, they had to print the DCT, send along the components, and call the ward to announce the arrival of the blood component.

4. Safety for professionals could also be compromised without an adequate traceability, and with the lack of legibility and details in some handwritten prescriptions.

All these problems lead to patient safety issues, unnecessary workload, and dissatisfaction for the professionals.

With the support of the Hemotherapy Commission and the commitment of the whole hospital leadership team, including the CEO and the CIO (Juan Luis Cruz), the head of the blood bank (Dr. José Luis Bueno) led the efforts to implement an ambitious roadmap with an approach based on the following key aspects:

- Focusing first on patient safety and reducing workloads, reducing costs as a consequence of the former
- Using IT as a tool for supporting redesigned business processes
- Partnering with all the stakeholders involved in the transfusion process, including not only clinicians and nurses, but also IT and its providers
- Considering it not as an isolated project but as a continuous improvement cycle (Plan-Do-Check-Act) based on data analysis coming from interoperable systems
- Thinking big, starting small

The project was named "OPINTRA" standing for "OPtimización INformática de la TRAnsfusión" (Informatic Transfusion Optimization) and was developed in several phases with these main milestones:

1. Implementation of a stand-alone transfusion safety system. As a quick win for patient safety, in 2013, the hospital implemented the system that allowed us to do the following:
   a. Verify blood samples collected against the patient using barcoded stickers and bracelets.
   b. Verify blood components against the corresponding blood samples at the blood bank and again at the bedside against the patient just before the start of the transfusion.
   c. Register electronically the possible adverse events observed by the nurse.
2. The project implied a change in management and training for more than 800 nurses. The commitment of many nurses, acting as the project ambassadors and elaborating short formative videos, proved itself as a key aspect for the success of this phase.
3. Development of a clinical consensus guide about the utilization criteria for blood components by the Hemotherapy Commission. Published in 2016, it set some key aspects of the transfusion process, such as the appropriate hemoglobin levels for red blood cell (RBC) transfusion and a subtle key change in the prescription process, consisting of implementing a single unit blood transfusion with every prescription.
4. Implementation of a new and tailored blood bank management system, called e-Blue, fully integrated with the rest of the systems involved in

the transfusion process, including the Electronic Patient Record (EPR) (Selene from Cerner), analyzers, and the transfusion safety system previously implemented. e-Blue, which was co-created by the different stakeholders involved from the beginning, included these main features:

a. Integrated with the EPR, so:

  i. The final transfusion report (DCT) was automatically generated.

  ii. The system could evaluate if there was a valid blood sample for every patient and automatically cancel the nurse order generated with the electronic prescription, saving unnecessary nursing work and needle sticks for patients.

b. Fully designed to support the reengineered process, so:

  i. Every prescription (request) was linked to only one blood component.

  ii. The information required by doctors while prescribing was reduced dramatically.

  iii. Availability of lab results along with the prescription so that the blood bank could decide if the transfusion was well-indicated according to the new hospital-approved guide and also allowing later analysis about adherence to the guide on a per-doctor basis.

  iv. Integrations with the rest of the systems involved and new user interfaces allowed nurses and doctors in the wards and the blood bank staff to know in real time the status of every prescription, in a simple way, without needing phone calls.

  v. A new user interface allowed the implementation of an active quarantine hemovigilance program so that a nurse could review every transfusion result and record undetected or not reported and adverse events.

5. Outcomes measurement and further process improvement. While not totally concluded in its first iteration, we have measured some results so far:

a. Errors avoided: There were 55,636 blood sample collections (SC) and 79,395 blood component transfusions (BCT) verifications using the transfusion safety system during the four years of the study. We found 1,995 (3.59%) SC and 548 (0.69%) BCT incorrect verifications (near-misses). The near-miss rates were broadly different depending on the hospital ward, the user, or the time when the SC or BCT took place. The highest near-miss rates occurred during the night shift. Electronic transfusion safety systems are able to reduce the BCT errors and also detect the near-misses incidence. They also allow us to detect gaps so as to improve personnel re-training and reschedule not urgent transfusions in those wards where near-misses are frequent.

b. Better management of adverse events: In 2016, only 13 adverse transfusion reactions were reported with the previous passive hemovigilance system (6.1 incidents per 10,000 blood components transfused), while after carrying out the active hemovigilance process, the number of identified events was 102 (48.2/10,000). Therefore, the number of incidents detected using the new system was 7.9 times higher than those detected by the passive system. In 2017, the number of events detected was 143, corresponding to a rate of 74.8/10,000. This rate is 6.77 times higher than the average rate reported for the last seven years, which corresponds to 11.05 events per 10,000 transfusions. Thanks to the new systems, we could also detect late adverse events that occurred after finishing the transfusion, especially dyspnea or other lung-related adverse events (TACO/TRALI). In 2017, the average time for an adverse event to occur was 5.5 hours from the beginning of a transfusion. In conclusion, the system allowed us to increase the number of identified adverse events, especially those that occur after the end of transfusion. The increase does not mean a higher incidence of adverse events but a better detection of them and better care for our patients.

c. Blood component consumption reduction: After the implementation, we have experienced a 7.3% reduction in the consumption of red blood cells and a 34.6% reduction of plasma. Apart from implementing the single unit prescription policy, knowing the actual hemoglobin level used for transfusion in different settings will allow us to educate prescriptors proactively.

d. Avoiding unnecessary work and costs: Paper has been eliminated from the whole circuit, as well as duplicate blood sample collection and analysis at the lab. Less time fulfilling the final transfusion reports, re-labeling, or on phone calls has created additional savings, as well as avoiding the need to return empty bags to the blood bank by hospital porters (that is about 15,000 displacements per year!).

Working on this project has generated new, innovative ideas for the future, such as using analytics, including natural language processing and machine learning techniques to automatize, finding influencing variables, and making predictions. As an example, these technologies would allow us to do the following:

■ Identify variables influencing transfusion prescriptions, errors, and adverse events so we can establish a risk score for patients and act proactively.

■ Automatize detection and information collection regarding adverse events buried in the free text of the EMR.

To summarize, these are in our view the key takeaways extracted from our experience:

■ Identify an actual and relevant problem and solve it.

■ Focus first on patient safety and second on reducing workloads and dissatisfaction for the professionals. Reducing costs will be a consequence of the former.

■ Get your leadership team involved.

■ Use IT as a tool for supporting redesigned business processes, not imposing them.

■ Involve all the stakeholders from the beginning of the project; partnering with IT and its providers is also critical.

■ Measure before and after, and take actions based on data.

■ Despite how we manage specific projects, innovating is a process, and as such, it should be managed and improved incrementally.

■ Think big, start small. Results take time.

Notes

1 Prioritizing the Patient Experience. www.jonespr.net/prioritizing-the-patient-experience

2 NPS Benchmark. 2022. What Is a Good Net Promoter Score? Retently. www.retently.com/blog/good-net-promoter-score/

3 Trends in Healthcare Payments Annual Report. InstaMed. www.instamed.com/white-papers/trends-in-healthcare-payments-annual-report/

4 Healthcare Leaders Survey Report: Achieving a 360 Degree View of the Patient. Verato. https://verato.com/resources/sage-report/

5 National Poll on Healthy Aging. University of Michigan Institute for Healthcare Policy and Innovation. www.healthyagingpoll.org/reports-more/poll-extras/pandemic-disruptions-mean-many-older-adults-still-havent-gotten-needed

6 Arnetz BB, Goetz CM, et al. Enhancing healthcare efficiency to achieve the Quadruple Aim: an exploratory study. BMC Res Notes. 2020; 13: 362. www.ncbi.nlm.nih.gov/pmc/articles/PMC7393915/

Chapter 3

Create Roadmaps

Develop a plan for the functions required to innovate and encourage effective communication between functional experts for strategic clarity.

There is an unsubstantiated fear that plans and order run counter to the innovation spirit. Effective roadmaps actually serve as beacons or markers that help innovators navigate their way without being distracted and thrown off course. Plans do not stifle innovation but rather provide necessary guardrails to ensure focus and completion. Too many great ideas were never realized as resources and passion dwindled from an unnecessarily long journey.

* * *

Virtual Vision

Steve Hess

Virtual health is the future, right? But, what does that really mean? The pandemic forced all of us to think differently about how we deliver care, and we all scaled virtual visits because of the worldwide shutdown. But, virtual visits are just a new way, potentially a more efficient way, of delivering individual care and care interventions between a patient and a provider.

Virtual visits are not new, not particularly innovative, and still require lots of clinicians and lots of clinicians with availability. What if there were a way to have a virtual care team watch over all kinds of patients, with all kinds of

DOI: 10.4324/9781003372608-3

conditions, in the hospital, out of the hospital, and everywhere in between? That would be interesting.

It was January 2020, and like many others, UCHealth had created capability for virtual visits as a new way of delivering care. Like previous months, we had 1,000 completed virtual visits in the month across primary care, specialty care, and urgent care. Just two months later with a global pandemic becoming an escalating reality, UCHealth completed 77,000 virtual visits.

Like many, we responded to the crisis. Unlike many, we just scaled an existing solution that had already been in place and fully integrated into the electronic health record (EHR).

While the shift to virtual health felt like an overnight phenomenon, the journey at UCHealth started many years prior.

Find the Burning Platform

In 2016, UCHealth was planning the build of three new smaller (<100 beds) community hospitals. We knew these new hospitals would need resources to conduct patient monitoring and surveillance. We knew that we would need telemetry technicians to conduct cardiac monitoring. We knew we would need sitters to watch over patients at risk of falling. And at the same time, we were contemplating the idea of a virtual ICU surveillance program for the new ICUs within the new hospitals. We were also a fairly new health system. In 2012, three existing health systems (with five larger hospitals) came together to form UCHealth. We knew we wanted to use IT, our enterprise EHR, and virtual health to help bring our system together and serve the Rocky Mountain community and our patients who live there. So, the need to create services at newly constructed hospitals and the desire to align our new system became a burning platform to create the UCHealth Virtual Health Center (VHC) that could serve not only the new hospitals but all hospitals under the UCHealth umbrella.

Create the Roadmap and Walk into the Future

The vision was clear and the organization appetite to think differently and divert spending from the new hospitals to a centralized capability was there. We knew we wanted to create an offsite area that would be staffed 24x7 with technicians, nurses, and providers who would use technology, intelligence, and their clinical expertise and serve as a virtual safety net for the patients in beds across UCHealth. But, we also knew we didn't want to be limited by the walls of the hospitals. So, while the new hospitals were being constructed, we started with a direct-to-consumer, urgent care–like focus and created a 24x7 capability to allow our patients to get care no matter what time of day from the comfort of their home. Since we were walking into the future, we stood up the capability using existing resources from existing care areas, such as freestanding emergency rooms. In September 2016, our first month of operation, we had 11 total visits. Yes, 11 visits in an entire month. While volume was extremely low, we, and our patients, saw the value, and we knew we were onto something. At the early stage of the pandemic, in March 2020, we completed 4,200 virtual urgent care visits with 45% of the visits because of COVID. Our patients were scared, didn't know where to turn, and found our Virtual Urgent Care offering to be a lifeline. Over two years later, our patients are still turning to us and this offering. In October 2022, we completed 3,600 visits.

We no longer have to borrow resources from other clinical areas and have technicians and providers who are now dedicated to virtual urgent care, and our Virtual Urgent Care center is the second busiest urgent care across the 21 urgent cares within UCHealth. *Virtual urgent care* is here to stay.

Deliver Economies of Scale

In 2017, we created and staffed a centralized team of technicians who can conduct cardiac monitoring under the VHC umbrella. If we had put telemetry technicians in each of our newly constructed hospitals, we would have likely had one technician in each hospital monitor ten patients simultaneously. By centralizing our cardiac monitoring, each of our telemetry technicians can monitor 42 patients simultaneously. By bringing resources from across the UCHealth system, and by creating a centralized construct, we created economies of scale and were able to absorb the volume of the three new hospitals with existing resources. Today, we monitor 550 patients each day across the UCHealth hospitals via *Virtual Telemetry*.

In parallel with the cardiac monitoring capability, we created and staffed a centralized team of technicians who, with the help of cameras in our patient rooms, watch over our patients who are at risk of falling. Most hospitals have to staff sitters who physically must be with the patient, in the room, and make sure they don't fall. This one-on-one sitter staffing model is difficult, expensive, and unsustainable. By centralizing our tele-sitters and using audio and video technology effectively, each of our tele-sitters can watch over 12 patients simultaneously. Today, we surveil 72 patients at risk of falling each day across the UCHealth hospitals with our *Virtual Safety View* program.

At the same time, we created and staffed a centralized 24x7 ICU-trained nursing role to provide ICU-level surveillance within our medical/surgical (non-ICU) units. We integrated wearable and standard bedside monitor device data, fed that data into algorithms sitting on top of the EHR, and the nurse in the VHC could watch over our patients and help identify potential issues and then collaborate with the bedside clinicians to change treatment plans as deemed appropriate. Back in 2017, our Virtual ICU team could watch over about 100 patients simultaneously given the relatively new algorithms and intelligence we had. Today, each *Virtual ICU* nurse watches over about 1,000 patients, and we now have several nurses taking care of 2,000 ED and inpatients.

Be Opportunistic

The vision and roadmap were playing out successfully. But, frankly, we knew we were only scratching the surface of what we could do. For a capability like the UCHealth VHC, it was still very novel, not understood by most, and the change management lift was significant across both the academic hospital and the community hospitals. The return on investment was real, the economies of scale were real, and the ability to use technology, device integration, and intelligence was real (and getting better). But, the real clinical impact was still a bit unknown. Then, along came a renewed focus on sepsis and the quality goal of reducing sepsis mortality. Like many organizations, we wanted to get better at identifying and eradicating sepsis within our inpatient population. Like many organizations, we were implementing EHR-based alerts to serve up to the bedside nurses and providers to get them to intervene when there was indication of sepsis. And, like many organizations, we were swamped with false positives, and the alert fatigue was real, and we weren't moving the needle. So, we created our *Virtual Sepsis* process where the VHC nurses (and providers) get the EHR-based alerts, do the clinical deep-dive review, and, in collaboration with the bedside team, can call the sepsis alert, initiate the sepsis bundle, and get the patient to a better place quickly. We eliminated over one million bedside alerts (most of which were false positives), reduced time to antibiotics by 31 minutes, and reduced time to fluids by 56 minutes. Whereas a typical nursing unit at UCHealth would see a sepsis alert once every 60 days, our VHC clinicians were seeing sepsis four times per day. It isn't hard to see who will be better at identifying and responding to sepsis.

Success Begets Success

Since the implementation of Virtual Sepsis, we have now created the ability to detect any patient deterioration, not just sepsis, Our *Virtual Deterioration* capability is now live and producing incredible early results. Rapid response events are up, which means we are catching patient deterioration earlier, and therefore, Code Blues are down (up to 70% decrease in our hospitals). Patients are staying where they are and don't need to be transferred to a higher level of care. More importantly, our patients are being discharged home, which, obviously, is the ultimate goal.

We are now implementing a similar capability to surveil our patient population at risk of developing pressure injuries, using a centralized nurse to look across the enterprise using the EHR and EHR-based algorithms, identify patients at risk, and then collaborate with bedside teams to inter-vene with patient turning, skin checks, and other interventions. *Virtual Wound Care* was born. Similarly, we are implementing the same approach for patients on ventilators who need to be monitored and weaned off of the device. And, now we have *Virtual Respiratory Therapy.* With the early results we are seeing, and the fact that we are standing up many of these capabilities without adding additional resources, the UCHealth VHC is often the first tool pulled out of the toolbox when trying to solve challeng-ing problems across our system. The VHC is a critical part of our quality and safety agenda.

Knock Down the Hospital Walls

We didn't ignore those patients outside the walls of our hospital either. During the pandemic, when we needed to decant the hospital to take care of more acute patients, our VHC was home for a program where we identi-fied stable patients in the hospital who needed one or two more days of monitoring. Instead of doing that monitoring within the hospital, we dis-charged those patients, often with oxygen, but also with a device that mea-sured blood oxygen levels (SpO2). The wearable data would be ingested into the tools used by our VHC, and we would monitor those patients for up to eight days after discharge to ensure they stayed safe. We called this our *COVID RPM* program, and we took care of more than 660 patients with this offering. Even after the pandemic has settled out to be normal course of business, we still use this capability to ease capacity constraints within our hospitals.

We also introduced a program that was more of a traditional remote patient monitoring (RPM) program with the initial focus on patients diagnosed with diabetes. To date, we have enrolled and/or graduated 250+ patients with an average HbA1c decrease of 26%. Our *Diabetes Home and Remote Care* capability is a foundation for several other diseases, such as congestive heart failure (CHF).

Appreciate the Journey

Today, the VHC combines ten services all interwoven together to create a patient monitoring and surveillance engine that serves UCHealth extremely effectively. It is truly a virtual care team. It is staffed 24x7 by technicians, nurses, advanced practice providers, and intensivists, and each role is working top of license and use each other to deliver seamless and comprehensive technology-enabled care. Each day, the VHC conducts 4,100 interventions across more than 800 unique patients within and outside our hospitals. Like everything else in healthcare, our success is a combination of people, process, and tools. The real gold is the incredible clinicians within the VHC and at the bedside led by incredible clinical leaders in the VHC who are working hard to keep our patients safe. These leaders are clinical experts who are visionaries, creative, masters of change management, and incredible partners with our IT team. None of what has been accomplished has been easy. It takes financial resources, patience, and the will and desire to manage change and win the skeptics over and deal with the complexity of a large health system. Looking back at the journey, it all started with a vision and a roadmap. We used a burning platform to sell the dream, and we figured out a way to walk into the future. The walk becomes a jog, and the jog becomes a run. Some of what we are doing now was envisioned; some of what we have done has been opportunistic. What the VHC looks like today will be very different than tomorrow, but the fact that we have the VHC opens a lot of doors that couldn't be opened without it.

* * *

Governance and Innovation

Kristin Myers

As an academic medical center that is part of a large integrated health delivery system in New York City, Mount Sinai Health System has a strong history and growing reputation of innovation, whether it is in our research teams, our physicians, our venture partner's team, our commercialization team, the Institute for Precision Medicine, the data science teams, or within our own technology department. Innovation includes not only the introduction of new products, services, or business models, but also embedding it into our organization as part of our culture. There are so many innovative efforts across the health system, but I will focus on technology innovation.

As CIO and Dean for Digital and Information Technology, I drive the transformation of digital and technology as an organization with the goals of driving organizational agility, optimizing technology operations, and enabling technology innovation. My scope of responsibilities span digital enablement, enterprise data and analytics, clinical data science, clinical and nursing informatics, intelligent automation, artificial intelligence (AI)/machine learning, cybersecurity, service delivery, enterprise applications, cloud/ infrastructure, interoperability, and Project Management Office (PMO). As Dean for Digital and Information Technology, I also have direct oversight over academic IT, research informatics, and scientific computing. My role enables me to partner with key stakeholders across the organization to drive, manage, and support technology innovation.

Over the years, it has become very apparent that technology innovation has been exponentially growing within our health system. The expansion of innovation is beneficial for the organization but has also brought forth challenges as well. One of the largest challenges we face is understanding all of these innovative technology pilots and efforts that our departments have developed, ensuring alignment with our health system strategy, and determining how we operationalize and scale this into the enterprise workflows for our caregivers, patients, and employees across the entire health system. At the same time, we need to have the appropriate governance in place to make data-driven decisions to ensure alignment with our strategic goals with a focus on a simple and seamless end-to-end experience and journey for our caregivers, patients, and employees.

Governance is not typically associated with innovation, however without it, the innovative pilots and efforts across the health system may never gain traction, or its value may never be fully realized as large enterprise rollouts of major functions such as enterprise resource planning, revenue cycle, or electronic medical records take precedence. One of my key objectives is to enable governance to support technology innovation by leveraging existing forums to evaluate these innovative solutions; providing the resources, technology, and infrastructure to support the efforts; and also scaling the pilots to health system solutions. A more agile governance structure is leveraged to help establish controls and processes that can expedite decision-making when appropriate and not impede work.

I will first outline our application portfolio strategy and management approach that lays the technology foundation and how governance plays a part. I will then provide some examples around how the digital and technology department has partnered with our business stakeholders to explore new innovative models and technologies and how governance ultimately can support, operationalize, and scale innovative solutions.

Application Portfolio Strategy and Management

Understanding the overall application strategy that provides a roadmap and direction for each major clinical and business function is critical. Our health system leans toward a platform-centric strategy for major functional areas such as electronic medical record/revenue cycle, enterprise resource planning, and service delivery. However, there are major gaps in functionality for these enterprise systems that need to be addressed to improve workflows, operational efficiency, and patient and employee experience. It is important to recognize those gaps, understand from the enterprise vendor if this is on their development roadmap in the next one to two years, identify innovative solutions, implement and evaluate the solution, and then integrate into the enterprise workflow and technology environment.

The first step to defining the application strategy is to develop an application inventory. It has taken a large effort to understand the application inventory, review and vet the application strategy with clinical and operational stakeholders, and develop the associated application roadmaps by function. It has provided a deeper level of transparency and clarity around the current state of our application portfolio and environment, the future direction of application and technology investments, and the key capability gaps to be addressed through technology innovation. This could include piloting a new application (large majority), leveraging an existing application in a different way, process automation, AI/machine learning models and algorithms, etc. Lastly and perhaps more importantly is having the ability and resources to evolve pilots to full-scale implementations at the health system level.

The key to this process and alignment has been governance to ensure there is an agreement between technology, operations, business, and clinical teams on future direction with clear visibility into new business needs and new health trends. The need for governance was driven by lack of communication and agreement on the target state strategy—technology teams had minimal visibility into business needs, and our business stakeholders did not fully understand the enterprise application strategy and the potential impact these new solutions and technologies have to our existing technology environment. This led to applications being piloted or implemented that were not in alignment with the enterprise application strategy, redundant applications and solutions of similar capabilities being implemented at different sites, inconsistent technology and operational workflows across the health system, and overall wasted effort, resources, and costs.

Application governance has drastically reduced these issues. The objectives of application governance include making application investment decisions that are aligned to our business and technology strategy, reducing cybersecurity risk, increasing transparency of the application portfolio, and providing a consistent application strategy for our health system.

Through a formal intake process, a weekly Application Steering Group forum, and escalation paths if needed, there is increased visibility into all new applications and pilots, formal processes to evaluate pilots, minimization of application redundancy and duplicative efforts, and consensus and data-driven decision making on new application investments and pilots.

This allows for early planning and forecasting of investments, resources, and support needed to scale application pilots across the health system if successful.

Application pilots are a good way to understand the application's capabilities and how well it addresses our needs. They typically last three to six months, have limited integration with our health system, are concentrated in a specific area, require minimal resources, and are generally no cost or low cost during this pilot period. The key to evaluating a pilot is to establish metrics and expected outcomes at the beginning of the pilot and then assess whether they are met at the end of the pilot. Usability, efficiency, ease of use, integration complexity, and cost are also important factors to consider. At the end of the pilot, governance again plays a part here where impacted parties and decision-makers come together to make a collective decision on whether to move forward.

Once the decision is made to move forward, our teams need to come together to architect the overall solution of how to incorporate the technology into the environment and existing workflows. From a technology perspective, the overall solution needs to take into account the application, data, security, infrastructure, and integration aspects. From a workflow perspective, the application needs to be embedded into the workflow for a seamless, simple, and consistent experience. The solution should not be requiring duplicate documentation from our end users, fragmenting the caregiver or patient's experience through accessing different applications, or being inconsistent at different hospital sites within our health system.

These issues ultimately lead to failure of adoption and decreased satisfaction from our patients and employees. All solutions should strive to be consumer-centric, experience-led, and equitable with a digital first mentality.

Telehealth

Before the COVID-19 pandemic, we conducted a few years of telehealth experimentation. This is a good example of our innovation lifecycle where we worked with a number of partners to learn and determine how we can evaluate pilots, make governance decisions around target state application strategy, decommission redundant applications, and scale the appropriate applications for successful implementation across the health system as part of an integrated clinical workflow.

Prior to commercial reimbursement for telehealth, we surveyed that we had more than 20 telehealth solutions across the health system. Many were low-cost or even free that were being used for specialty consults or direct to patients for care. This spurred the need to establish telehealth governance across the health system that was led by a business sponsor with participation and support from technology, operational, and clinical departments. The business sponsor was appointed to review and vet the technology and the experiences of the physicians and to identify organizational and regulatory barriers to successful telehealth. Over the next few years, some of these technology companies went out of business, the regulatory/reimbursement models changed, and we identified a successful telehealth pilot to integrate within the clinical workflow of the enterprise electronic medical record. Once this was finalized, capital funding was approved for telehealth in this model to be implemented across the health system.

As this is a top priority for our health system, we have been gaining a lot of traction with telehealth initiatives and improving the experience for our patients. The increase in telehealth is inevitable with 15%–20% of all ambulatory care volume now being telehealth. We have defined and implemented an integrated holistic experience for virtual urgent care and scheduled health and are also creating new innovative models such as virtual primary care.

We look to continue expanding on telehealth initiatives and measuring the benefits by leveraging our existing governance structures.

Robotic Process Automation

Another example of an innovative technology pilot that eventually scaled to an enterprise solution and program is robotic process automation (RPA). This is a technology that configures software (a bot) to mimic the human interactions with systems and users to execute repeatable and rules-based processes. Our technology leadership identified RPA as an area that should be explored to automate work in finance, supply chain, and other operational areas. After discussing with the operations team in ambulatory care, a pilot was identified to automate the current manual process performed by patient coordinators for loading new patient packet data that is received through emails. The opportunity was to reduce the amount of time for the patient coordinator to collect data and manually load it into the application, thus improving data loading accuracy and reducing errors. We also identified another pilot within the technology organization to reduce manual efforts on provisioning. As a result of the success of both pilots, an enterprise RPA program and governance was established that is co-sponsored by me and the chief financial officer. Since then, we have confirmed the enterprise RPA technology solution and expanded the team with approximately 50 processes automated and approximately 70,000 FTE hours saved this year alone. We continue to proactively work with different departments to identify and execute on potential automation opportunities.

Clinical Data Science

Our clinical data science function sits within the technology organization with a close partnership with other technology teams and clinical teams to ensure synergy. The goals of the clinical data science team are to support clinical departments through developing and operationalizing machine learning-based decision support applications and to optimize the use of machine learning and AI to improve quality, safety, and the patient experience. Data scientists have been recruited to develop data models in sepsis and other clinical conditions to be able to develop predictive tools embedded into our electronic medical records platform so we can deliver these tools at the point of care for our physicians. Our electronic medical records vendor has a number of predictive models they have developed, however, rather than only using the data models the Mount Sinai data scientists develop or the vendors' models, we developed a governance approach that is complementary. Requests for models are reviewed against the vendor's and roadmap and then prioritized and resourced for all technology teams.

In the last two years, a new AI governance committee has also been established in response to the growing concern and issues around AI ethics. The purpose of this layer of governance is to ensure the development, evaluation, validation, and use of AI is ethical so that our algorithms are clinically safe, effective, and equitable. This committee is focused on evaluating the ethics aspects of internally developed and external commercial AI and machine learning algorithms. The clinical data science team's machine learning and AI algorithms are also reviewed at this committee to support our AI ethics goals.

This is an example of collaboration and partnership between different technology teams and governance groups that support innovation in alignment with clinical goals and objectives.

Digital Transformation—Patient Experience and Access

The digitization of every aspect of our lives combined with increasingly empowered patients is shaping consumers' expectations. This has led our health system to seek to redesign our care models to deliver high-quality and seamless care and experiences at every touchpoint. The digital transformation journey is a top priority for our health system to reimagine the digital strategy around how experience-led, equity-based digital capabilities will improve our employee and patient experience, bridge the digital divide, improve access, and reduce cost of care. This will be done through engaging patients and employees in new ways, expanding the business models, and enabling a digital omni-channel strategy and operating model. Creating a seamless patient journey experience that will connect, inform, coordinate, and communicate referrals from outside the hospital with a transfer center, transparency to patients with regard to future bills, having a single access center for entry for new appointments, and future scheduling are key. Patient access also includes how organizations connect, engage, serve, and ultimately retain patients throughout their journey.

The digital transformation program is a close partnership, engagement, and governance between the digital experience team (that sits within the technology organization) and operations teams. A joint digital vision was agreed upon, and a new digital governance structure was established. Digital governance will enable the oversight of digital initiatives and technologies and will interact with other governance committees to remain integrated with the broader strategy. This new governance is extremely crucial to reduce duplicative efforts, technologies, and costs and for the organization to share a consistent digital vision. The governance structure also includes the Digital Equity Committee, which will review and monitor key digital capabilities and initiatives to ensure that it is equitable and aligned to the broader enterprise diversity, equity, and inclusion initiatives. Prior to this enterprise effort, there was an overabundance of digital pilots and point solutions across different departments, which led to a fragmented patient experience, an overly complex technology environment, and duplicative and inconsistent workflows for our caregivers.

This governance structure is crucial to the review and approval of the end-to-end experience design lifecycle as it provides a framework to assess and measure new digital capabilities. The design starts with the patients at the forefront where we conduct patient focus groups, interviews, and surveys to understand pain points and areas of opportunity. Employee feedback is also critical to understand the gaps and improvement areas for our frontline staff and operations teams. The new digital capabilities are prioritized based on impact to patient and employee experience, return on investment, digital equity impact, complexity, etc. For example, having multi-language options for the MyMountSinai mobile app is a high priority due to feedback received from patients and its alignment with our health system diversity, equity, and inclusion objectives. The value and impact of these new digital capabilities and experience will feed into our overall digital business case where we track patient revenue increases, referral increases, leakage reduction, patient satisfaction, employee satisfaction, and more.

This is an example of where a large-scale innovation program is managed end-to-end through enterprise governance to optimize the use of resources and investments.

Conclusion

We live in a "VUCA" world where there is volatility, uncertainty, complexity, and ambiguity. As the rate of innovation continues to grow through new industry trends, new business models, changes in consumer behavior, and technology disruptions, it will only become more imperative to have a close partnership with business and clinical operations and an appropriate governance structure to help manage and achieve the maximum value from the innovation investments. Governance does not need to be a heavy-handed bureaucratic process but instead can be an agile guiding body and framework to ensure transparency, strategic alignment, data-driven decision-making, collaboration, and better results.

Technology departments need to collaborate with all stakeholders and respond with vision, understanding, clarity, and agility. We need to proactively sense opportunities and follow through with collaboration and action with a focus on simplification. The risk of not expanding, supporting, and governing innovation with our partners leads to missed opportunities, pilots that do not get scaled to enterprise solutions, and challenges of attrition due to low employee motivation.

Fostering an innovative culture is also critical as part of the innovation journey. An innovative culture and mindset will help the organization more quickly and better adapt to the increasing change around us. This also includes talent management to develop human-centric skills to effectively serve and collaborate in this environment. Skills such as conflict management, cross functional collaboration and decision making, customer service, teamwork, and emotional intelligence are becoming essential for those who want careers in technology innovation.

Innovation continues to evolve, therefore, we as an organization need to as well. With an appropriate amount of governance, organizations can maximize their innovation investments and help drive a supportive environment to enable the growth of innovation.

* * *

Finding the Art of the Possible

Josh Sol

Empowering organizations to create an emotional connection by leveraging technology speaks to the heart of how digital transformation impacts the healthcare industry. Billions of dollars are poured into new technologies that strive to create ease, precision, and efficiency and are focused on the delivery of how health care is provided. It's important to recognize that managing multiple efforts, spreading resources too thin, and not having a clear understanding of an organization's vision to adopt a digital experience will convolute the intentions of what a process should be accomplishing. The art and simplicity of knowing your path is most certainly the North Star but is rout with "boulders to push up hill," "ships to steer," "political land mines to avoid," and "clunky tech with manual work arounds."

Leaders are often asked, "Why can't the path for healthcare be simpler? Why does healthcare technology not work like other industries? Why are we just now, at the tail end of a pandemic, realizing the power of digital technology?" As a user, a patient, and an implementer in the middle of this ever-changing landscape of endless choice, healthcare inequity, payor reimbursement reduction, an aging population, and staffing shortages, we need to understand the technology we are asking patients and clinicians to adopt and ensure it actually works within their organizational and operational structure. Often, good, bad, or indifferent, the user experience is compared to the understanding of the last technology they utilized. If an airline app was easy to use for purchasing a ticket and/or self-check-in, the patient expectation is that this same experience should translate to scheduling an appointment with their provider with an expected positive outcome. The initial experience plays a pivotal role in the connections consumers make, particularly healthcare, and is what drives the patient experience through their journey.

Why does this matter when we discuss creating journey maps? Simplicity, while creating tangible, actionable efforts, is key to enabling transformation. Healthcare delivery is complex, and digital enablers should continue to put into perspective that we are working with consumers and patients who are in or could be in very vulnerable states. We as healthcare professionals, no matter the industry focus, must recognize that we are supporting people during their care journey and the importance of empowering those who are caring for them. This should be the primary mission of digital enablers.

With those ideals in the backdrop, the key question industry leaders are facing is how we can transform an organization that may face staffing complications, technology debt, outside competition, internal naysayers, and budgetary constraints that put limits on what they can do. I contend that while we face many burdens in the industry, we can do this thing called transformation through small incremental steps, stages, phases, whatever a group wants to call it, but the goal is to get it started! A great saying, how do you eat an elephant? One bite at a time.

To understand the metaphorical elephant, the organization must honestly assess their current positioning and their future goals. Here is where the mapping begins. It is critical to understand where an organization is in order to plan how the organization can move forward and is the primary reason for mapping. This is where the "Art of the Possible" lives. "With what we have . . . what can we do?" Mapping begins, quite simply, with discussion. The idea is to get a detailed understanding of the organization's workflow both from the user's side and the technology side. In the example of a patient journey map, it could be from the time the individual is a consumer, shopping/searching how to find the right provider and/or level of care that complements their need. The technology can be running a million different algorithms or processes to make the patient's experience simple, but it must be unassuming to the patient. Imagine a duck on the water, smooth, stream-lined, and steady on the surface, but those little legs are certainly churning fast underneath to get the duck from point A to point B.

The imagery of the duck is how the journey map should feel. Lots of technical motions with if/then logic questions for the current state process, as well as how the user interacts with the technology. This initial progression allows for an understanding of where the pain points are, which can vary from organization to organization and highlights areas of opportunities and are where organizations look to establish their path for differentiation in a personalized experience. It's not pleasant to shine a light on a process, to expose areas that need to be addressed. In fact, it can be rather personal to the leader or the organization. However, in order to transform, the opportunities must be exposed. In my experience, the leader likely knows where the strengths and weakness are—they wouldn't be in a leadership role if they didn't—but these exposures help to build a plan to drive forward. Again, I go back to, how do you eat an elephant?

Moving forward, beyond the initial interviews and the build stage of the journey map, we look to formalize areas of opportunity. This draws out the pain points into opportunities on the same current state visual aid tool. The opportunities can be seen from within the organization, within the industry, and often from outside industry. The perception is that healthcare lives about 20 years behind other industries. We still leverage fax and pagers as some of the primary communication tools. We continue to move forward year after year, but somehow, we still live with the same technologies. The "Art of the Possible" lives within this stage. We look at the pain points, current process improvements, the industry innovators, and outside industry leaders, and we meld those items together onto the pain points/opportunities map. This new map drives the solutions and future themes the organization may want to focus on through their digital journey.

As the digital mapping assessment moves on, the themes become the mission, the mission drives the strategy, and the strategy drives the transformation. The themes are built from a combination of the assessment interviews, the understanding of the need for change, and the development of the opportunities. The combination of pain points and opportunities ultimately becomes the solution. While IT is about technology, technology is not successful unless people use it in a way that makes sense and can leverage it to gain efficiencies in their day-to-day experiences. Through the journey mapping practice, the organization may realize technology isn't the right answer. Technology is meant to simplify a user's life and should tie together the experiences of all users. The solutions that are created out of the initial assessment of the pain points and opportunities should do just that—create a simplified, streamlined, industry best practice opportunity that fits an organization's need.

This arrangement of simplicity, industry best practice, and leveraging existing technologies in the organization will contribute to the future state, "North Star" journey. This is where the magic happens! The dream of how to get to that "North Star," the development of the strategy that leads to differentiation in the industry. The transformation of the experience begins with this future state journey map. Leaders in the organization, if the map is done with passion to change how healthcare is delivered, will see the "Art of the Possible."

A future state map should have a combination of immediate impact with future opportunities. The impact and value proposition of those opportunities is where the next stage of magic happens. Each solution must be given a value. This is a combination of many different factors. By quantifying the solutions, the value to impact a model can be built. This model is what leads the organization to "the how."

The journey map is a masterful approach to understanding process and designing the future state, but eating an elephant should be about taking one bite at a time. The modeling creates those bites and allows for the organization to understand where the most value is with the type of impact they are looking to achieve. If it's low-hanging fruit with high impact, do it, and do it quickly. Why wouldn't you? Innovation and transformation successes will be incremental but long standing. These small wins begin the process of the flywheel concept. A mentor of mine used the flywheel concept to illustrate those small incremental wins and build momentum, and with that continuous momentum, the flywheel moves on its own.

The flywheel metaphor speaks to incremental change that supports the organization that can transform the mindsets toward innovation and that leads to positive change within an organization. The flywheel starts with the value to impact work effort. This takes what was a journey map and makes it tangible. I continue to use metaphors to illustrate that eating an elephant is about tangible bites. Building a digital strategy, or actualizing a future state is all about tangible actions. Those actions create momentum that builds and builds and becomes continuous. The road map that is produced to get to the future state sets the stage for governance discussions, active planning and budgeting, resource allocation, and prioritization of effort. These business drivers often look to a return on investment, otherwise why would an organization go through this exercise?

Return on investment, particularly in cases of innovation, is not always about the money—a former colleague of mine would always say, "No Money, No Mission." Technology is often a budgeted expense meant to support the clinicians and business units. The ROI for a technology has many different flavors, and each organization has an appetite for the different return on investment an application can bring. In the case of a frictionless check-in experience, the ROI may be on ease of use or throughput of patients when they arrive at the organization, or it could be as dramatic as rearranging your waiting areas to be smaller because the patients are already prepped and ready to be seen. This discipline to ROI, to the success of each tangible bite of the transformation, will make a difference in how the organization manages new ideas and innovation in the long run.

The journey map sets the stage, while the road map sets the strategy and has the ROI as its success differentiator. This process of planning to define the road map can be leveraged for processes that intertwine technology and the user. Seeking out strategies for the latest and greatest opportunities in health care, like supporting inequities of care, social determinants of health, hospital at home strategies, virtual and rural health, inpatient differentiation, and efficiencies, can all be supported by a disciplined look at process tailored through this journey mapping effort. I think the Field of Dreams illustrated it the best: "If you build, they will come." If you can highlight the areas of opportunity, build the map, find the prospects, focus on the drivers, drive toward the target, and just get going, an organization will transform to its next phases of healthcare delivery. Tangible actions can make a dream into reality.

* * *

Open, Connected Environment Drives Healthcare Transformation

Gary Johnson

This essay describes the transformative innovation deployed within the University of Kentucky Healthcare (UKHC) system that created a dynamic financial catalyst that expanded patient access and improved patient outcomes. This innovation was built upon the framework of ambulatory pharmacy services and grew to represent 70% of "income from operations" for the health system while reducing readmissions by 40% and expanding access for insured and uninsured patients.

Health system administrators have been trained to view pharmacy services as a cost center where variable costs can be depressed. This traditional paradigm is not applicable to ambulatory pharmacy services where growing drug expenditures can generate new income streams. In contrast, outsourced or shrinking ambulatory pharmacy expenditures is a strategy of retreat that atrophies organizational growth opportunities.

UKHC faced these same cultural challenges; however, with innovative ambulatory pharmacy models, UKHC inverted this cost-containment paradigm. This inversion increased ambulatory pharmacy drug costs by 500%, resulting in 900% margin increases through care models that expand patient access, with these financial returns reinvested into UKHC to continue the provision of services for all patients, including self-pay and indigent patients. This essay reviews guided efforts to optimize ambulatory pharmacy services, which include the following: (1) Optimization of contract pharmacy services via the federal 340B drug pricing program; (2) optimization and expansion of retail pharmacy services; (3) creation of a prescription delivery service for patients being discharged from the hospital, which is integrated with medication reconciliation-related processes; (4) optimization of employee prescription benefits; and (5) creation of a robust specialty pharmacy in collaboration with subspecialty clinics. The topics reviewed in this essay include how patient access was expanded, how the fiscal impact was measured, the barriers circumvented and how lessons learned were applied to other departments.

Compliance

UKHC spent 12 months focusing efforts on 340B program compliance as a foundation before moving ambulatory pharmacy initiatives forward. UKHC now employs multiple analysts to audit various aspects of the 340B program on a weekly basis. Having a regular cadence of compliance reporting allows for accurate monitoring of the program. For stewards of the 340B program, a process of baseline assessment and continual reassessment of compliance is essential as it allows the organization to continue accomplishing its mission while fulfilling the intent of the 340B program. This framework of compliance provides scaffolding for the UKHC ambulatory care pharmacy strategy, which includes three distinct operational units: (1) Retail pharmacy services, (2) contract pharmacy services and (3) specialty pharmacy services.

Retail Pharmacy Services

In the summer of 2011, at the beginning of this innovation strategy, UKHC had one flagship retail pharmacy that served multiple clinics as well as a satellite retail pharmacy in an adjacent building that served the UK student population. However, the lack of 340B-qualifying software prevented either of those retail operations from functioning as an "open pharmacy," which precluded the provision of services to many patients with prescriptions from non-UKHC providers; many of these patients were employees of UKHC. Until that point in time, hospital leadership was focused on cutting costs in the satellite retail pharmacy, which had lower volumes than the flagship retail pharmacy. The pharmacy department demonstrated that this satellite retail pharmacy generated a modest annual operating margin (i.e., net revenue minus operating expenses). This exercise generated interest in measuring the performance of the flagship retail pharmacy, changed the traditional cost-centric paradigm, and initiated a culture of revenue generation as opposed to cost containment. A pro forma for expanding hours in the flagship retail pharmacy, which entailed increased labor costs but would expand patient access and increase the net operating margin, was subsequently submitted. These experiences facilitated the expansion of five additional retail pharmacies throughout the health system during the next five years.

The flagship pharmacy in the clinic building was across the street from the main hospital, which presented a challenge for patients with discharge prescriptions as well as those with emergency department (ED), same-day surgery and employee prescriptions. To improve services for these patient populations, a business plan for a hospital lobby-based retail pharmacy service with a concierge medication delivery component, termed Meds-2-Beds, for discharge-related prescriptions was created. The importance of this service is ensuring that discharge prescriptions are dispensed to prevent unnecessary hospital readmissions. The added importance of providing this service for a safety net hospital (SNP) is that the discharge process—a critical point in transitions of care (TOC) for any patient—is especially critical for those patients who, when they reach their home destination, find that they cannot afford their medication from a local or mail-order pharmacy.

An SNP is prepared to perform a financial assessment of these patients, provide vouchers needed to dispense discharge medications and serve as a resource for processing prescription refills.

This lobby-based discharge pharmacy generated more income than expenses within the first month of operations. The high frequency of Meds-2-Beds service utilization generated visibility for retail pharmacy operations, which presented unanticipated challenges. One year later a similar hospital lobby retail pharmacy offering Meds-2-Beds services was launched at a sister community-based hospital, with similar success. More recently, UKHC opened a large clinic building at an off-campus location where retail pharmacy services were initiated in the building lobby. Within a year, expansion plans aimed at accommodating the overwhelming needs of these clinic patients were submitted.

Technology

UKHC replaced the pharmacy management system (PMS) in the retail space with a new vendor product, and the pharmacy department partnered with this vendor to create a real-time solution to the 340B-related problem. This new solution combined a PMS with a 340B-qualifying system that determined eligibility in real-time. The PMS interfaces directly with the patient registration file and the eligible-provider file, which eliminates the time lag created by the previous retrospective reporting method. At the point of prescription adjudication, the prescription is flagged for either 340B or WAC billing, which enables more cost-efficient product selection. The new PMS also eliminated the need for a third-party administrator to split-bill the order. Instead, dispensed prescriptions accumulated in the PMS and separate replenishment orders (at 340B and WAC prices, respectively) were placed with the wholesaler. This solution was later launched within the UKHC specialty pharmacy and proved critical, as the differing financial implications of WAC and 340B-dispensed prescriptions are more pronounced with specialty medications.

Contract Pharmacy

These services include the health system partnering with external pharmacies to repatriate revenue back into the health system. Traditionally, these partnerships include large chain pharmacies, such as Walgreen's, Walmart and CVS.

However, this presented another innovative opportunity to create contract pharmacy models with non-traditional partners, such as home infusion pharmacies and specialty, mail-order pharmacies located in other states. This process of innovation began in the area of greatest need for UKHC: Home infusion pharmacy services. UKHC does not have a home infusion pharmacy, and extending 340B participation into this arena was vital given that a large percentage of the patients are discharged with home infusion orders. Since home infusions are billed as medical benefit claims and not as pharmacy benefit claims, UKHC developed a non-traditional contract pharmacy model with a 340B split-billing vendor and a local home infusion pharmacy. UKHC also included provisions in the home infusion contract pharmacy agreement to support indigent and self-pay patients who account for roughly 15% of UKHC's home infusion patients. These contract provisions allowed UKHC to extend its SNP role by providing home infusion pharmacy services to patients who would have difficulty accessing home-based services. UKHC structured this relationship such that 340B program related savings would directly pay for medication vouchers generated by social services staff across the UKHC enterprise, which allowed UKHC to directly leverage 340B program savings to extend to care to the underserved.

Subsequently, in the summer of 2013, as the specialty pharmacy pipeline continued to grow, it was apparent that many of UKHC's patients were being forced to fill their specialty prescriptions through mail-order specialty pharmacies outside Kentucky. In the absence of a UKHC specialty pharmacy, UKHC reached out to the largest mail-order specialty pharmacy that was serving UKHC patients to discuss a contract pharmacy relationship. This concept was unfamiliar. This specialty pharmacy partner and both parties struggled with the many facets of implementation. Ultimately, this model resembled the home infusion contract pharmacy model, accommodating both medical and prescription benefit claims. The contract pharmacy model was activated in the spring of 2014, and this experience provided a foundation for the eventual development of a Utilization Review Accreditation Commission (URAC)–accredited specialty pharmacy at UKHC, which was moved into a dedicated space in February 2016.

Specialty Pharmacy

Specialty pharmacy services provided by UKHC in 2012 were limited to oncology and non-oncology infusions, a hemophilia treatment center, a state-sponsored AIDS Drug Assistance Program and an average of a dozen self-administered specialty medications dispensed weekly through UKHC retail pharmacies. In collaboration with eight large subspecialty clinics, it was determined that the total potential revenue from self-administered specialty prescriptions within these clinics was more than $200 million annually. The existing retail pharmacy infrastructure was providing specialty prescriptions for less than 1% of the patient's prescribed specialty medications. Furthermore, the same review uncovered dozens of medication access issues, fragmented workflow and general patient and clinic dissatisfaction with external specialty pharmacies.

In a validation of the quality of the clinical and technical services provided by the specialty pharmacy, specialty pharmacy accreditation was obtained in 2015, and a dedicated 5,000-square-foot site for the operation was secured in February 2016. This space houses a call center, dispensing and management personnel, a mail-order operation and all related inventory. Accreditation also led to improved access to restricted medications and improved contracting opportunities with Medicaid and commercial payers. Specialty pharmacy services in 2017 included care for patients with nine primary diseases and dispensing of an average of more than 2,500 prescriptions monthly, which included specialty and related supportive medications.

Because of the unique role UKHC fulfills as both an SNP and a specialty pharmacy service provider, there is an intrinsic investment in patient care by the pharmacy staff. UKHC pharmacists from the specialty pharmacy meet one-on-one with these patients in clinics to build strong relationships and ensure appropriate access to these medications. Specific pharmacy employees serve as access points for patients who cannot afford their specialty medications—or even the high co-payments often associated with specialty medications.

UKHC then pivoted the contract pharmacy dynamic and leveraged the UKHC contract pharmacy as a contract pharmacy partner for health systems throughout the state. This model allowed many small- to medium-sized health systems to generate margins from specialty pharmacy prescriptions that otherwise would have been impossible.

Financial Measures

During the initial year of ambulatory care pharmacy development, effort was expended developing business plans, pro-forma documents and return on investment (ROI) analyses to demonstrate opportunities. This work later translated into developing analytic platforms for measuring performance and

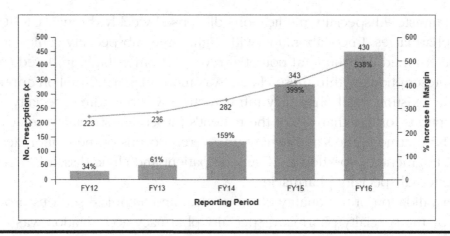

Figure 3.1 Performance of UK HealthCare ambulatory care pharmacy operations, by annual prescription volume (trend line) and annual percentage increase in operating margin relative to the fiscal year 2011 (FY11) margin (bars).

growth in each strategic ambulatory care pharmacy program. Over a five-year period, prescription volumes increased from 223,000 to 450,000 (93% growth), while the associated operating margin grew 538% by 2016 from the baseline of 2011 and 900% by 2018 from the baseline of 2011.

Discussion

This experience demonstrated the value of ambulatory care pharmacy services. However, this experience also demonstrated the complexity of operating a retail business within a highly regulated acute care environment. The complexity of these models extends beyond operations and reflects cultural challenges, which include the need for hospital executives to pivot from acute care to ambulatory care pharmacy services. This pivot is challenging for executives with only acute care experience who have been trained to assess pharmacy operations as a cost center as opposed to a revenue center. This dynamic requires pharmacy leaders to educate and incrementally demonstrate value associated with ambulatory care pharmacy models.

Another entrenched paradigm is resistance to the propagation of outpatient pharmacy solutions, which include the operating system, automation and tools to ensure regulatory compliance, such as software for managing 340B program complexities. The drug expense for many ambulatory care pharmacy operations nearly equals the drug expense for inpatient operations. As both inpatient and ambulatory care pharmacy costs grow, organizational leaders, including pharmacy leaders, might unflinchingly incur million-dollar software expenditures for inpatient operations, such as an electronic medical record, but grimace at ambulatory care pharmacy infrastructure costs, such as the cost of an operating system that exceeds $50,000 per year. Inpatient pharmacy investments dwarf investment in ambulatory care pharmacy operations. Ironically, in some cases, pharmacy directors who are continually pressured to lower drug costs have minimal capacity to influence utilization that would allow opportunities for revenue creation. Ambulatory care pharmacy services often compete directly against large retail pharmacy chains situated in and around medical center campuses.

Another challenge to ambulatory care pharmacy development is the lack of competency in that practice area within the pharmacy leadership. Pharmacy directors and managers have long focused on inpatient operations while perhaps failing to recognize the extent to which ambulatory care pharmacy has evolved in both complexity and opportunity. The regulatory environment, specifically within a 340B program model, is particularly onerous and breeds an abundance of caution. This caution and lack of understanding can intimidate pharmacy leaders and retard the development of legitimate opportunities.

A growing chasm between inpatient and outpatient pharmacy operations is being bridged with TOC models. These models, which blur traditional boundaries, include pharmacist-based clinic interventions in targeted populations, such as patients undergoing transplantation, as well as medication reconciliation at admission and discharge counseling about the bedside delivery of discharge prescriptions, which improve HCAHPS scores and rates of completion of medication reconciliation at discharge. Through these models, the discharged patient can be triaged into the ambulatory care pharmacy model to receive prescription services, including specialty prescriptions, as well as various forms of infusion services in the home setting or infusion center.

With the success of the larger UKHC ambulatory care pharmacy initiative, pharmacy services administrators found that opportunities to support the organizational mission in the form of expanded patient care are vast. UKHC expanded pharmacy services by adding more than 30 newly hired clinical pharmacists to inpatient service lines and ambulatory care clinics, such as the cancer center, cardiology clinics and the transplant clinic. Additionally, with the creation of the specialty pharmacy, clinic-based pharmacist roles geared toward enhancing care for patients with ten specialty diseases have been created and filled. The pharmacy department has implemented TOC technician roles that involve 24/7 staffing in the ED and other service lines with a commitment to thorough and accurate medication histories at admission. Inpatient pharmacy services leaders have adopted the Better Outcomes Through Optimizing Safe Transitions (BOOST) methodology, whereby inpatient pharmacy teams have expanded coverage to offer thorough medication reconciliation at admission with the assistance of medication histories captured by TOC technicians, as well as traditional high-quality inpatient clinical services, discharge medication reconciliation and medication counseling for patients who choose to have their medications dispensed through the Meds-2-Beds program. Also, as mentioned earlier, additional roles were created to assist indigent and self-pay patients with medication access. Without the success of the ambulatory care pharmacy initiative, these additional inpatient services and resources for patients would not be possible.

Conclusion

Opportunities for ambulatory care pharmacy services to provide valuable patient care and improve fiscal performance exist. UKHC expanded patient access to medications while creating a financially viable model that optimized existing retail pharmacy operations and opened additional retail pharmacies, implemented discharge prescription delivery services, implemented contract pharmacy services and opened an accredited specialty pharmacy.

* * *

Big Rocks Along Our Roadmap

Will Conaway

Where might healthcare be in the next five years? No person or organization can predict the direction with 100% accuracy, but evaluating the current state of healthcare and factoring in trends can provide an anticipated path forward, even if it's murky at times. Using mission and vision as trail markers, we can continue on that path with the mindset to improve healthcare for all.

Let's look at several "big rocks" we can expect to encounter on our hike.

Government

There will continue to be government involvement in new healthcare-related rules and regulations, and some will result from evolving consumer expectations. For example, federal mandates on interoperability have been driven by consumers' expectations about their Episode of Care and also led to statutes around healthcare providers giving patients access to all their electronic medical record information and doing so quickly and at no cost.

Population Shifts

Expect more changes with government-run healthcare programs as our population transforms significantly.

> *The year 2030 marks a demographic turning point for the United States. Beginning that year, all baby boomers will be older than 65. This will expand the size of the older population so that one in every five Americans is projected to be retirement age. Later that decade, by 2034, we project that older adults will outnumber children for the first time in U.S. history.*[1]

Healthcare Becoming a Business

We will see more new entrants into healthcare as a business. The pool of entrants will not be limited to well-known names such as Microsoft, Amazon, Google, and Apple, who can create latent needs for healthcare with their products. The pool will include others, too, who understand and acknowledge a truism that many industries and sectors have recognized for years: Technology trends focus heavily on consumer empowerment. New technological advancements and innovations are limitless. In the past few years, there have been many patient-centered advancements with data analytics, automation, artificial intelligence, machine learning, cloud, XaaS, virtual reality, and wearables. With each innovation, another one is generated, as all in technology push each other. Innovative technologies will continue to provide products and services to assist its base with better research. Healthcare availability with new locations for care will empower patients to have a more active role in their care.

Getting to the Scenic View

The items we have noted thus far on our hike down the roadmap path are just a few of the "big rocks." Being an agile leader and having flexibility in your healthcare roadmap is a must to get over, under, or around these rocks and use them to your advantage.

"Plans are Nothing; Planning is Everything." ~ President Dwight D. Eisenhower

After years of being on the other side of the table, I now work in a partnership role in healthcare. I will not use the term *vendor* since no longer can a company grow in healthcare without a partnership mindset. Whether you are a partner or provider-based system, your healthcare roadmap must be able to deliver on the customer experience. You must offer services that differentiate you and define your price transparency strategies. You must also use unconventional methodologies that are disruptive. The best companies understand the concepts of "the innovator's dilemma" and will no longer make the mistake of missing out on new waves of innovation.

As a healthcare partner, I have to identify trends in healthcare quickly and adroitly and proactively anticipate customers' needs when creating healthcare technology roadmaps. No matter what side of the table you are on, there is one constant: If you are in healthcare technology, you are part of the patient's Episode of Care. This is a heavy responsibility and can never be taken lightly.

As a healthcare system leader, you may ask why you should be highly concerned with changes in healthcare and invest the time to develop an enhanced roadmap. A great example of the "why" is the COVID-19 pandemic, which caused unprecedented, rapid changes in healthcare in everything from diagnosing to treatment delivery. For example,

> During the pandemic, 18% of primary care visits were virtual—soaring from a mere 1% in 2018. The share of virtual primary care has settled around 12% at the end of 2021, much higher than pre-pandemic, according to the report. Young healthy patients will continue to embrace digital and virtual healthcare services. Bain (Bain and Company) predicts that virtual health could climb to 20% of market penetration by 2030.[2]

"The First Responsibility of a Leader Is to Define Reality. The Last Is to Say Thank You. In Between, the Leader Is a Servant." ~Max de Pree

Where do you start to create a roadmap? Start with the mindset that your strategy is not only what you do, but what you decide not to do. To establish a good healthcare technology roadmap, one must be able to define things. Defining may seem an easy step; however, I continue to find that the biggest threat to strategic roadmap success is the inability of critical stakeholders to define things the same way. It is nearly impossible to set clear objectives as part of your planning process if you cannot clearly articulate what you are trying to attain.

Make no mistake, multiple challenges will make it critical to be able to define your roadmap. As with most industries, healthcare finds its strategic roadmaps based upon assumptions derived from the extrapolation of historical data. As we know, the past can be a good predictor of the future, but creating a roadmap needs to deal with the future, and the future is laced with VUCA.

> *Volatile—change is rapid and unpredictable in its nature and extent. Uncertain—the present is unclear, and the future is uncertain. Complex—many different, interconnected factors come into play, with the potential to cause chaos and confusion. Ambiguous—there is a lack of clarity or awareness about situations.*[3]

Sometimes in the planning process, goals will become strategies, and strategies will become goals. Once a strategy becomes a goal, it sometimes spins off its own strategies. Developing goals from strategies or vice versa is not a strange metamorphosis. In establishing an organizational strategy, one should note that strategies apply to the "how" of achieving the goals and are not the goals themselves. Anytime a strategy is being developed, concerns regarding its effect on an organization must be addressed and evaluated. When formulating strategies, consider these basic rules:

1. Strategic roadmaps will determine the direction of things. Your strategic roadmap should not change with each meeting.
2. Strategies are part of your objectives, but do not confuse the two for the same thing since your objectives are the expected results.

3. Your strategies must be communicated lucidly, and you will need confirmation that they are understood.
4. Make sure you have the right people to deliver and a single point of contact. When everyone is in charge, no one is in charge.
5. Allow time for your strategies to work, but do not make them so rigid that there is no flexibility.

"You Can't Manage What You Can't Measure." ~ W. Edwards Deming

Metrics set the magnitude for your roadmaps. A good metric does three things. It lets you know where you have been, where you are today, and if you are tracking toward success. It is vital to assign the appropriate type of measurement to your initiatives. For example, a mature company addressing learning and growth, business processes, customers, and finance may want to implement a balanced scorecard. But if your organization is looking for more concrete results, including a tool to communicate what you must accomplish and your required milestones, an OKR (objectives and key results) methodology is more suitable.

Control is assessing and regulating both work in progress and work completed. No strategic plan is better than the ability to control and measure it. Measurement, evaluation, and corrective action should be a continuous process. After-the-fact evaluations and measurements have a place in control hierarchies. However, if evaluations and measurements indicate corrective action is required, damage can be minimized by in-process corrections.

Here is a tip that will save you much trouble. As a Lean Six Sigma Blackbelt, many healthcare executives ask me how to make a metric that "makes things look good." The key is to measure the items that *can make things better*, not only those that make things look good.

"Those That Fail to Learn from History Are Doomed to Repeat It." ~Winston Churchill

It would be impossible to isolate or itemize how an organization undermines or supports creativity. Oddly, if a person states that their organization undermines creative processes, maybe that person is not applying sufficient job-related creativity. I have found that most people will contribute creatively to their environment. Unfortunately, an organization's structure and foundation can limit a person's work-related creativity level.

Fostering consistent innovation requires policies not just encouraging it but also allowing it to be applied to bring about further innovation and transformation that's in tune with an organization's strategic realities. Applied creativity must exhibit value to customers while still being at a price-point they are willing to pay. With this thought, know there is also a time to discard today's creative innovations. A properly structured organization will continuously create new and improved replacements that supersede yesterday's best ideas.

Regarding mission statements: A mission statement should be a living document subject to review and revision as the business matures. It should generate plans, goals, creative ideas, and strategies, and as the business grows, the mission statement should complement this growth.

If a mission statement has not evolved or has not been reviewed for a long time, it could limit the innovative capabilities of the organization's people. Reviewing an organization's mission statement is worthwhile. It can help define the employees' position within a firm more clearly and demonstrate how each person can contribute to accomplishing a firm's mission.

The goal for all organizations should be to have every employee understand the firm's mission as well as their own role in the organization's success. When this happens, they become more creative, productive, and loyal because the mission now relates directly to them. Often, firms miss tremendous opportunities by underestimating (or ignoring) their employees' ability to contribute to success. These missed opportunities hamper innovation.

Of course, there are costs and challenges related to innovation. In general, budgets set goals, evaluate results, and improve decision-making by planning, communicating, and coordinating the various components or disciplines within a firm. Conversely, when preparing operating and capital budgets, most managers give little or no thought to employee creativity or how to fund creative efforts.

"Excellence Is Not a Destination, It Is a Continuous Journey That Never Ends." ~ Brian Tracy

Anyone can point out a problem, but your best healthcare leaders find solutions. Once strategies are formulated, the next logical step is implementing the plans and applying those strategies.

Henry Mintzberg, in his book *The Rise and Fall of Strategic Planning*, made some very pointed remarks regarding strategy implementation. Mintzberg wrote the following:

> *Strategy formulation should be the open-ended, divergent process in which imagination can flourish in the creation of new strategies, while implementation should be the close-ended convergent one in which these strategies are subjected to constraints of operationalism. But because of planning's need for formalization, ironically it is formulation that becomes tightly constrained, while implementation provides the freedom to decompose, elaborate and rationalize down the ever-widening organizational hierarchy. The consequences, as we shall see, are that, planning formulation lost its creative potential while implementation provides great powers of control.*[4]

To be effective, the implementation process of strategy management requires all strategies be broken down into manageable sub-strategies in order to do the following:

■ Solve a strategic planning bottleneck.
■ Break the issue into components.
■ Solve the component parts.

Only then can the elements be reassembled. After many years of working with students at Cornell University, it pains me to write that the neat and clean model observed in textbooks is not a reflection of reality. The parts do not always mesh as diagrammed. Roadmaps have a dichotomy of formulation and implementation. They can sometimes only identify specific people in an organization to create strategies, oftentimes creating a schism between the "Strategists" and the "Tacticians."

Take the example of leaders committing to battle and then pointing to others to participate. With rare exceptions, the underlying reason for your healthcare roadmap's failure is neither in the creation nor the implementation but in the detachment of the two.

"It Is Not the Strongest of the Species That Survives, Nor the Most Intelligent, but the One Most Responsive to Change." ~Charles Darwin

The future for healthcare is bright. We are seeing more functional experts driving innovations that will be game changers for patient outcomes and experience, promote equity and lower costs, and ultimately create a paradigm shift in how you receive your Episode of Care.

Healthcare leaders must retain accountability and responsibility, which are vital to their roadmaps. And do not underestimate the importance of good communication. Plans are only as good as their controls, measurements, and communications; if these aren't up to snuff, neither will your plan be.

You must be able to communicate and hire for functional expertise. You must also recognize trends in talent shortages and devise innovative strategies to adjust accordingly.

According to an article by LexisNexis, "The statistics are ominous." The article continues:

Nearly one in five of America's healthcare workers left the field during the COVID-19 crisis and more say they are planning to exit soon.

The American Hospital Association reports that at least 23% of U.S. hospitals have experienced a critical staffing shortage since February 2020.

With more than half a million nurses expected to retire by the end of 2022, the U.S. Bureau of Labor Statistics projects the need for 1.1 million new registered nurses.

The U.S. faces a projected national shortage of more than 3 million low-wage healthcare workers over the next five years.

The Association of American Medical College forecasts a shortage of nearly 140,000 physicians by 2033.[5]

Additionally, talent is a scarce resource. Yet, in many organizations, when present, this resource remains underutilized. A primary reason for the underutilization of creative talent is unenlightened management that does not understand how to identify or use talented people.

Underused or misused talented people often become frustrated and unproductive. To rectify this problem, leaders must learn how to utilize creative and talented people by placing them in positions encouraging creativity and resultant innovations. Creative processes can be chaotic for some organizations, yet those that encourage and support creativity and innovation will be the survivors.

A Few Closing Words on Technology

The last couple of years have been difficult. The healthcare industry hasn't just been another industry affected by the "great resignation"; it could be seen as the industry that has experienced the "great redesign," with technology playing a key role.

Technology has created new realities in which healthcare organizations must diversify and grow in order to survive. As an industry, we are seeing many changes and from different players than before. The right diversification allows an organization to generate a business in which the performance capacity is equal to that of the top performers. Technology is a component of diversification that can be used to assist an organization to branch into a different area. It is more problematic to construct market diversification with ordinary technology. In order for technology-based diversification to be successful, the plan must be specific, distinct, central, and not incidental to the organization's healthcare services.

I am confident that through technology our healthcare leaders and providers will have insights as never before that will allow for better decision-making that will advance healthcare and benefit all. However, understanding the origin and the source of information contributes to its value and the amount of consideration this information deserves. Even in this technological age, it is difficult to minimize the human element when making a decision.

Automating the information gathering processes will prove advantageous. For decisions that require more than minimal human interaction, information technology could be applied to assist human decision makers.

I am confident information will change established cultures and sub-cultures within healthcare. This breaking down of existing cultures will be accompanied by the establishment of new cultures and new members joining these new cultures where healthcare will be more available and in all geographic locations, which in turn will enhance patient outcomes.

* * *

Roadmaps

Bill Russell

The single biggest problem in communication is the illusion that it has taken place.

—George Bernard Shaw

A story has main characters, a plot, a villain and a resolution to a challenge that the characters must overcome. The development of innovation roadmaps tells a story of a potential future and how a system might organize to get there, but it's not the whole story. We tell incomplete stories, and in so doing, we fail to identify the right challenge and to connect with the very people that are the main characters in the story. In short, we fail to inspire action.

In the fall of 2010, I began work on a five-year plan to turn around a struggling IT shop for a 16-hospital system as their CIO. We did all the expected activities. We did a listening tour, conducted surveys even organized collaborative sessions to dream about the future and consider the various paths to get to the place that would meet our mission while staying true to our values. We explored the potential of technology and the role and relevance of the IT organization. We believed that a compelling narrative was needed to capture the urgency and potential of technology to address the particular needs of healthcare at this particular moment in time.

We developed a roadmap for transformational change that we believed as an organization was necessary to be prepared for an uncertain future. This roadmap was defined by two core initiatives: A move to the cloud and a data-driven strategy. The move to the cloud represented our belief that a new level of agility was going to be required in healthcare to be relevant in a rapidly changing landscape. Given what is happening in 2018 with new partnerships and business models being introduced almost weekly, that assumption seems to have been correct. A data-driven strategy was recognizing that data is one of the most important assets a company has in the digital world, and we believed that healthcare was in the midst of a digital revolution. Data would be the fuel that drove the triple aim in a digital economy. Now we needed a way to communicate the strategy to the stakeholders.

My leadership team did what most would do at this point: we developed a PowerPoint deck, and we started presenting our collaboratively developed vision of the future. It is widely recognized today that while PowerPoint may be a good tool, it promotes walls of text and bulleted lists, which become one of the worst forms of communication. When we were done creating the deck, we presented it to a few groups and realized it was uninspiring and didn't motivate action. The message didn't connect with people on an emotional level. If people are expected to act on something, they have to know and feel certain things before they will invest emotionally.

We stepped back and retooled our communication, but what follows is more about the mistakes we made and what I've learned since then. For those of us challenged with building a culture of innovation, what follows are some lessons for how you can take the elements of your roadmap, organization strategy and technology plan and weave them into something that can motivate people to action.

I would be remiss if I didn't mention that professional story tellers and designers are your friend. If it weren't for Andy Parham, Paul Bussmann, Jim Fisher and our design sessions, I would never have been able to tell our story.

People Think in Pictures

When I say "Red Ferrari" you immediately picture a red sports car, you don't think about the letters "R" "E" "D" that make up the word. Our brains are hardwired to translate words to pictures.

We had several slides that described the challenges that our health system faced in 2010. Every time we started with these slides, we noticed that people would reach for their phones to check their email. Spending too much time telling people what they already know is a hallmark of poor communication. We finally enlisted a team of creatives to help us with the process. The result was a single picture that described the state of our system.

In Jonathan Swift's book *Gulliver's Travels*, Gulliver the giant is tied down by a thousand tiny ropes by the people of Lilliput. The graphic designer told us that as we described the challenges facing healthcare, he couldn't help but see this image in his head. The next time we presented, we had one slide: A graphic with the image of Gulliver tied down by Lilliputians and no words on the slide. We talked about how healthcare was held back by 1,000 tiny ropes, payment models, culture, interoperability, application sprawl, communication and other factors. We talked about how it wasn't any one problem that held us back but many that work in conjunction to hold us back from being what we could be as a system. And no one reached for their phone.

A Good Narrative Invites People to Step Into the Story as the Hero

We tend to tell the story in a way that doesn't allow people to enter the story. We tell the story where we are the hero, the system is the hero and sometimes even the vendor is the hero. Ask yourself if this resonates. We are buying this new EHR that is going to improve quality, reduce friction and increase efficiency. No one would do this today, but we were reading these stories weekly 10 years ago. The EHR is the hero of this story. It's no wonder it suffers a reputation problem. The EHR stole the lead role in the story from the very people we know to be the heroes of the healthcare story.

How about this well-intentioned narrative: Our founders came to this community with nothing but their grit and desire to serve, and we stand on the shoulders of their great work. This is the start of a great narrative, as we will discuss later, but it needs to invite people into the story. Your founders would want the current generation to put on a cape and solve the greatest challenges of today. I'm not advocating for a rise of individual heroes in healthcare. In fact, our success has been found when we act more like NASA as a group with a shared commitment to a common mission than as comic book heroes. However, even when we lead with our heritage, we need to invite people to enter into the story and write the next chapter.

People want to be the hero of their own story. The system, vendor and technology are resources to help them overcome a great challenge. There are many examples of this, but the best one happened on May 25, 1961. President John F. Kennedy made this declaration before a joint session of Congress:

> *I believe that this nation should commit itself to achieving the goal, before this decade is out, of landing a man on the moon and returning him safely to the Earth.*

Notice that this was not a call specifically to NASA—it was a call to the whole nation. Everyone was invited to consider how they might be a part of the mission.

Every Story Needs a Villain That Needs to be Vanquished

President Kennedy presented this message in the context of good and evil. To open this section of the speech, he framed this work as part of a war between freedom and tyranny. Tyranny and oppression are good villains to the average person. We all know that a common enemy is a powerful unifying force. In this case, no one wanted to live under a dictatorship that was represented by communism during this time period. An important step for every organization is to identify the villain in your story.

There are many villains in healthcare in general. The obvious ones are the chronic diseases that threaten the lives of people in our community. In this battle, the heroes wear white coats and wage a war on behalf of the patient. The other villains that have been elevated in our national debate and communicated through the quadruple aim are access, cost, quality and experience.

Intermountain Healthcare CEO Marc Harrison, MD, addressed the villain of healthcare inequalities based on zip code in his comments at the 2018 JP Morgan conference. It has been written about many times that zip code, education and financial standing are more highly correlated to your health outcomes than genetic disposition to disease. Intermountain identified one of these zip codes and is partnering with Medicaid in Utah to take on a risk model that puts them on the hook for improving care in that geography. Health outcome inequalities is a good villain.

Deborah Proctor, retired CEO of St. Joseph Health, and Rod Hochman, MD, the CEO of Providence, cited a growing concern for mental health and the need for whole person care in the communities they serve on the west coast as one of the driving factors for their merger in 2016. They helped to fund the Well Being Trust in 2016 with an initial investment of $100 million to address mental health shortly following the merger. The previously glossed over problem of mental health in our communities is a good villain.

The role of leadership is to identify the right villains.

Aspirations Are Good, Calls to Action Are Better

Our aim is to provide the highest quality care to everyone in our community where both the patient and caregiver have a great experience at an affordable cost. This is a solid aspiration, and it helps to frame our activities, but it doesn't inspire and motivate to action. A more effective call to action will invite people to step in as the hero of the story and clearly identify the villain that needs to be overcome. Let's examine some calls to action from healthcare leaders.

Eliminate the waiting room and everything it represents.
David Feinberg, M.D. CEO
Formerly, Geisinger Health

Dr. Feinberg goes on to explain, "A waiting room means we're provider-centered—it means the doctor is the most important person and everyone is on their time. We build up inventory for that doctor—that is, the patients sitting in the waiting room." The villain in this story is the poor experience that healthcare provides to patients and their family. The symbol of that poor experience is the waiting room. The call-to-action invites staff to present work and budgets that eliminate every waiting room. Speaking on behalf of all patients, I hope they are successful.

> The future is healthcare with no address, where the patient is in charge.
>
> Stephen Klasko, M.D. CEO
> Formerly, Jefferson Health

This statement is more nuanced. The villain here is the limitations of the healthcare status quo. A health system that requires physical locations will be expensive and deliver a poor experience. This may also represent a culture that is unwilling to be progressive in its adoption of tele-strategies and the use of sensors for remote care. The call to action is to redesign healthcare to be a customer-facing organization and not a business-to-business transaction with the payer. Dr. Klasko's call to action invites the healthcare industry as a whole to vanquish the status quo and embrace new methods.

A good call to action, such as Drs. Feinberg's and Klasko's, paints a picture of a potential future: A hospital system without waiting rooms and a health system delivering care that isn't defined by its zip code. People see in pictures. Paint a compelling future picture of what healthcare can be to activate your roadmap.

Use Video to Validate the Vision

I've been particularly impressed by systems that have started to utilize video to tell a story. Kaiser Permanente has cut several videos that imagine a future. One example (https://youtu.be/NZm5gJikhgE) paints the picture of their care anywhere vision. One needs only to watch the video to imagine their place in making the vision a reality. The power of using video to communicate is self-evident; however, the value of the video is far greater than the $20k–$40k investment to create it. You can determine the level of alignment with a broad audience, including the medical staff and the community at large. The video acts as a tollgate in validating the vision with your constituents.

There are many examples of the use of video to imagine the future. There are two more that I will highlight; one is technology-related and the other was designed to impact the culture. United Healthcare (https://youtu.be/gEybaotgZgc) paints a picture of a future where sensors and technology keep you connected with a trusted care partner. The Cleveland Clinic's empathy video (https://youtu.be/pIGzPsfnpoc) is a powerful message on the role of empathy—a gentle reminder to be present, attentive and compassionate. Once you clearly see a possible future, investing in a video that helps people to see it clearly is money well-spent.

Finally, Connect Your Story to the Previous Chapters in the Book

The Sisters of Saint Joseph went into a town that needed education and healthcare with the intent of starting schools. A flu outbreak led them to meet the needs of the community that were most acute, and they started in healthcare. This lesson on its own is powerful, but the story doesn't end there. Some years later, the Sisters were the first health system in the region to utilize automobiles in the delivery of care. The automobile is the blockchain of its day. It's a compelling technology that everyone believed had an application for healthcare, but it needs creative people to find the right fit.

The automobile was the connection with our past that we were looking for. Our Sisters were innovative, adventurous and not afraid to try new things. We dedicated one slide to a black and white picture of our innovative Sisters standing next to their ambulance. The result was an unmistakable connection between our desire to utilize the technology of the day to serve the needs of the communities we operate in.

We tend to write a story as if it started with us. The reality is that most health systems have great formation stories. Connect the storylines.

We joke that every movie on the Hallmark channel has the same plotline with different characters, yet if I walk past a screen with one of their movies playing, I stop and watch. I try to identify who the characters are, who the villain is and the challenge that they must overcome to find love at the end of the movie. The existence of the Hallmark channel should teach us that there are certain storylines that work no matter how many times you use them. Hero meets challenge that is usually protected by a villain, a guide comes along to help them get past their personal limitations and overcome the challenge, which leads to a better future. I know how every Hallmark story is going to end, but the predictable nature doesn't detract from my interest—it enhances it.

If you consider yourself to be an innovative person with a technology bent, I wrote this chapter for you. This is my background and bent. What I learned in our process is that you have to invite people on the journey. People have to know their role, and the challenge has to be something that resonates with their being. Storytelling is the next step in taking your roadmaps, PowerPoints and project plans and stirring people to action. The more compelling your story, the more motivated people will be.

Reference

Miller, D., 2017. *Building a Storybrand: Clarify Your Message So Customers Will Listen*. HarperCollins Leadership, an Imprint of Harpercollins.

* * *

A Project Management Approach to Delivering and Optimizing Innovation

Gregory Skulmoski & John Walker

Introduction

If you want innovation, then the project management delivery approach will improve the probability of project success and subsequent innovation optimization. The key to successful projects is a healthy focus on risk and quality management by thoughtful alignment with ISO standards and other global best practices. We outline how to innovate with project management and use a point-of-care testing project to illustrate our approach. We include "microlearning" opportunities where the reader is further directed to resources online for more information.

Innovation Through Project Management

The project delivery approach has been the favored method to deliver innovation: The pyramids, space flight, and smartphone-connected pacemaker devices! Are there any viable alternatives to project management to bring a promising idea to life? We routinely organize project teams to deliver a product or service for use (Figure 3.2) and subsequent improvement. An innovation begins with a demand or opportunity to deliver value and is retired when the innovation no longer delivers value. Therefore, innovators

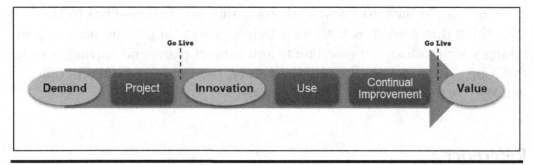

Figure 3.2   Innovation delivery approach.

will more likely succeed if they combine the right amount of project management and supporting standards to deliver innovative products and services.

Project Management Delivery Models

Modern project management began with the traditional delivery approach (also known as the waterfall approach for its predictable and linear path) and can be used for most projects (Figure 3.3). Adaptive project management approaches later emerged, like agile project management. Contemporary project managers tailor their delivery approach and use the best of both, ending up with hybrid project management.

Traditional Project Delivery Approach

The traditional project delivery approach is the standard way to implement most projects since it is a logical and proven process to deliver value

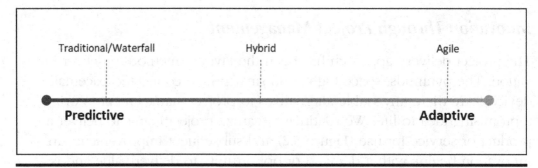

Figure 3.3   Project delivery continuum.

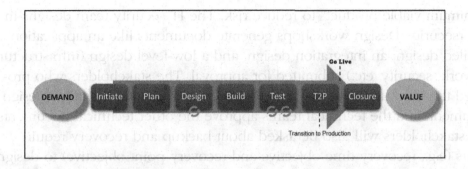

Figure 3.4   Project management delivery approach.

(Figure 3.4). The innovation may be documented in a business case and scheduled for implementation if successful. The innovation proposer (e.g., project sponsor) might be guided by comprehensive innovation frameworks to systematically detail the innovation (Keeley et al., 2013). Approved projects proceed to the initiation phase and are assigned to a project manager based on availability or specialized skillset (e.g., laboratory project implementation).

The project manager works with the innovation project sponsor in the initiation phase to convert the business case into a project charter: A one-page infographic with the project scope, rationale, high-level Gantt chart and budget, key risks and issues, and project manager authority to manage the project. The approved charter is used in the planning phase to develop a project plan. The project manager works collaboratively and iteratively with the stakeholders to develop the plan that is presented to a project change control committee for approval to proceed to the design phase (and later, we will work with the change authorization board to release the innovation to the end-user).

The project team follows the approved plan and collaboratively designs the innovation with the innovation-proposing stakeholders and technical teams (e.g., the clinical application specialists, infrastructure, end-user devices, IT security, networking, etc.). We ask, "Is there anyone missing?" in the planning and designing phases to ensure we have the right people. We may follow problem-solving approaches like lean six sigma or design thinking to improve the probability of developing a valued innovation. Sometimes elements of the design (e.g., a use case) might need to be tweaked, and the design may go through an iteration to find a suitable solution (Figure 3.4). Some sponsors elect to design the complete innovation but implement the "minimum viable product" to reduce risk. The IT security team designs-in cybersecurity. Design workshops generate documents like an application detailed design, an integration design, and a low-level design (infrastructure, network, security, etc.) submitted for approval. The stakeholders who proposed the innovation are asked to approve the application detailed design document, and the technical teams approve the other technical documents. The stakeholders will also be asked about backup and recovery requirements (e.g., recovery time objective and recovery point objective) to design a digital innovation with the right amount of availability.

An approved set of design documents guide the teams to develop the innovation in the build phase. However, most of the technical work is configuring rather than creating an application, perhaps following the DevSecOps approach and using a programming language like Python. Once the innovation is built, the project enters the test phase.

The purpose of the test phase is quality control (find and fix defects) on the product or service. We develop and follow a test strategy and plan that includes the testing scope, environment, types of tests, test data, test cases and scripts, defect management, etc. Testing is often executed in two steps, with the first tests without end-user involvement. The defects may be minor, like a spelling mistake, to more critical defects like integration challenges.

Sometimes defects trigger iterative problem solving (Figure 3.4). Once the defects are rectified and the system operates according to the approved design, the end-users are invited to test and approve the innovation. User acceptance testing is a critical milestone, and user approval allows the testing team to complete testing and submit the test report.

The team exits the test phase and enters the transition to production (T2P) phase (Figure 3.4) where the final preparations are completed to take the innovation live. Technical activities like testing the backup, penetration testing, disabling vendor passwords, etc., are completed. The project team may follow best practices like the ADDIE instructional design process, Bloom's Taxonomy scaffolding, and Kirkpatrick model of training evaluation to train the service desk and end-users (Skulmoski and Walker, 2023). Communications are prepared but not just for "challenging or worse" scenarios. Communication also needs to take a central place in the methodology and is a mainstay of stakeholder management. The project team implements the innovation, monitors its performance in the production environment, and gathers user feedback. After any necessary stabilization (e.g., rectifying minor defects), the project is closed out by completing final activities like a lessons-learned exercise, contract closure, and submitting the close-out report.

Microlearning

- What is a business case?
- What are Larry Keeley's 10 types of innovation?
- What is a project charter infographic? (and CV infographic?)
- What are the top ten project plan templates?
- What is a Gantt chart? (And, how to create a Gantt chart in PowerPoint?)
- How do design workshops work?
- What is a minimum viable product?
- How does lean six sigma work? (and search for free white belt courses?)
- How does design thinking work?
- What is the difference between ITIL response and resolution time?
- How do backup, recovery, and disaster recovery work?
- What is the difference between recovery time objective (RTO) and recovery point objective (RPO)?
- How does UAT testing work?
- What is the ADDIE instructional design process?
- What is Blooms Taxonomy scaffolding?
- What is the Kirkpatrick model of training evaluation?
- What are digital transformation projects' best practices?
- How does contract closure work?
- What is the Project Management Professional certification?
- What business lessons can we learn from goldilocks?

Adaptive Project Delivery Approach

There are times when the predictive nature of traditional project management (Figure 3.4) may falter, and an adaptive approach is more suitable to implement the innovation (Figure 3.5).

Adaptive project management is used when the end product or service is ambiguous, or the method is ambiguous. However, we still plan, design, build, test, deliver, and learn, but with iterative sprints (e.g., one month in duration). In each sprint, the team completes work prioritized in the product backlog. A benefit of the adaptive approach is that innovation value is delivered incrementally, and adjustments may be quickly made.

Once the innovation is delivered and used, the product or service owner collects recommendations for improvement and adds them to the product

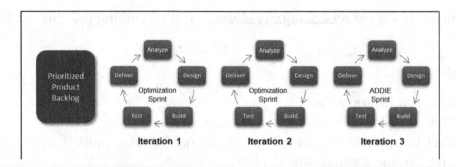

Figure 3.5 Optimization sprints.

backlog for implementation. Not only can we improve our innovation with optimization sprints (e.g., adding more functionality), but we can also use adaptive project management to deliver training with an ADDIE sprint (Skulmoski and Walker, 2023). ADDIE is generally accepted as best practice for instructional design where we Analyze, Design, Develop, Implement and Evaluate training; quality and project management are evident in ADDIE. Therefore, we can use adaptive project management to deliver further innovation and training.

Microlearning

- Search for the Scrum Guide (PDF, details the sprint process).
- Search for the Agile Manifesto values and principles.

Project Success Record

Unfortunately, delivering innovation is challenging; how challenging? Innovation projects can fail at a rate between 35%–50% (Castellion and Markham, 2013). The Project Management Institute reports that only 60% of projects are successful, and 22% are considered failures (PMI, 2017a). Why the poor results? They had project plans, schedules, and budgets. We believe project failure is often due to inadequate risk and quality management. We improve the probability of project and innovation success when we focus on quality and risk management and align with standards and frameworks that have risk and quality management at their core. We invite the subject matter experts proposing the innovation to ensure the ISO standard is represented in the design.

Quality Management: ABC

Quality management is at the heart of ISO standards (e.g., ISO 17523 Electronic Prescriptions, ISO 17523 Human Pathology or ISO 15066 Collaborative Robots). ISO standards guide the practitioner to safely provide their product or service to the planned level of quality. Indeed, the ISO 9001 Quality Management Standard is a widely adopted companion standard and includes a quality management process (Figure 3.6) we use in project management.

Quality management begins with planning (e.g., policies and procedures, roles and responsibilities descriptions, etc.). Quality assurance is fundamental to quality management, followed by quality control once the work is completed. We emphasize quality assurance to prevent quality "defects" (quality problems); we try to get things right the first time. Once we complete the work, we perform quality control to find and fix defects. Quality assurance is achieved through training, processes, checklists, and templates to guide our work. We improve the probability of finding and fixing defects by following a testing process with checklists to ensure completeness. Some projects produce a test report to document quality (e.g., a stage gate) before proceeding to the next phase.

PDCA Cycle

The iterative "plan-do-check-act" (PDCA) cycle guides users to include continual quality improvement in their work. Central to the PDCA cycle developed by Dr. Ed Deming is the iterative nature of quality management to improve our innovations. To conclude, quality management is central to project management and other standards to deliver the product or service to

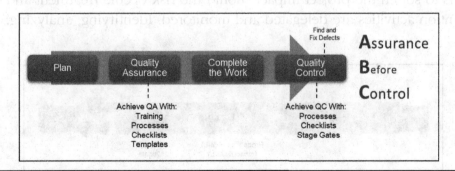

Figure 3.6   Quality management process.

the planned level of value. Following standards automatically brings about improved risk and quality management.

Microlearning
- Search for ISO standards relevant to your practice, and look for quality assurance and quality control elements.
- Search for Deming quotes such as "leadership is only 99% of the problem."
- Search for PDCA in innovation.

Risk Management

The single most crucial aspect of project management is managing project risks. Most projects have approved schedules and budgets, yet it is common for projects to be delivered late or over budget. Why? Risks become issues that interfere with delivering on time or on budget. The simple solution is to manage project risks early and regularly. We can rely on ISO 31000 Risk Management (Figure 3.7) for guidance.

Risk management starts with planning and includes artifacts like risk management policies and procedures, roles and responsibilities descriptions, and other governance items such as a risk and issue register. We identify risks that can impact our project and analyze the impact on the project if the risk occurs and the probability of occurrence. A lean analysis approach is to use High-Medium-Low for the probability of occurrence and impact on the project. This qualitative analysis is quick to complete and is suitable for most projects, most of the time.

The most severe risks (e.g., High/High) are prioritized for treatment, where we look for ways to prevent the risk. We also identify mitigation actions to soften the project impact should the risk occur. Treatment and prevention activities are delegated and monitored. Identifying, analyzing,

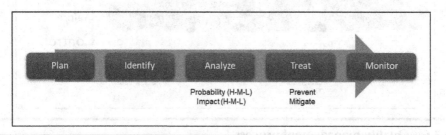

Figure 3.7   Risk management process.

treating, and monitoring risks is best practice. For example, during informal and formal discussions, we ask about the status of prevention and mitigation actions to confirm they are successfully in place and that the risk is being managed. The risk and issue register is the tool we use to document and manage our risks and issues. A risk becomes an issue when it is "realized" (e.g., when impacts to the project occur and are no longer possibilities). Effective risk management makes the difference between a successful and failed project.

Microlearning

- What is project risk management?
- How do risk registers work? (You can find risk register templates online.)
- What is the difference between a risk and an issue?
- What are the top digital transformation risks?

ITIL Service Management

Digital innovations carry the risk that the innovation may not fit very well in the organization's digital ecosystem, resulting in unmet expectations. Planning, implementing, supporting, and optimizing digital products and services have a long history of challenges. The Information Technology Infrastructure Library (ITIL) framework has emerged to guide organizations in managing technology (Figure 3.8).

A demand for a new digital innovation (e.g., smartphone-connected pacemaker devices) begins the ITIL process of introducing the innovation into the digital environment. That demand can be detailed in a business case, vetted, and prioritized by a clinical or business operations committee. A project is initiated (Figure 3.4), and the product or service is delivered to

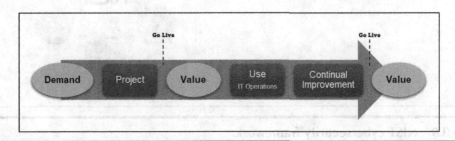

Figure 3.8 ITIL service value system (simplified).

the end-user, who determines its value. The innovation is used and subsequently improved with optimization sprints (Figure 3.5) based on the quality improvement PDCA Cycle. The innovation is used and optimized until it no longer delivers value.

Microlearning

■ What is the difference between ITIL 3 and 4?

NIST Cybersecurity Framework

Digital innovations need protection from cyber threats. The National Institute of Standards and Technology (NIST) Cybersecurity Framework outlines the five functions of cybersecurity (Figure 3.9).

Cybersecurity begins with an "identify" planning phase to prioritize which assets to protect. Measures are implemented to protect those assets (e.g., security orchestration, automation, and response software), followed by monitoring the digital ecosystem to detect threats. The IT security team responds to mitigate and recover when a threat is detected. Quality improvement is also part of the NIST cybersecurity framework.

There are critical points to consider: First, cybersecurity activity occurs within the operations environment. Second, we improve cybersecurity

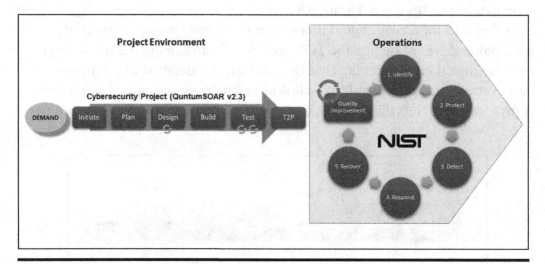

Figure 3.9   NIST cybersecurity framework.

capabilities through projects; we cannot maintain cybersecurity readiness by purchasing it online. In Figure 3.8, a cybersecurity project has been initiated and is in the T2P phase. There is a monitoring period after go-live to stabilize the application if required (illustrated with a vertical, dotted line). Third, cybersecurity threats are increasing and will continue to increase for the foreseeable future resulting in mandatory cybersecurity compliance for critical infrastructure digital innovations.

Microlearning

- Download the NIST Cybersecurity Framework.
- What is NIST role-based and general awareness training?

CASE STUDY: POINT OF CARE TESTING

Greg Skulmoski (co-author) was the project manager responsible to automate, integrate, and extend point of care testing (e.g., blood gases analysis) and used project management aligned with standards to deliver this innovation. A key benefit of testing and analysis at the patient rather than in the laboratory is reduced turnaround time. Cleveland Clinic Abu Dhabi is a HIMSS Stage 7 (Electronic Medical Adoption Model) hospital with a strong point-of-care testing (POCT) service.

Initiation

Jonathan Harris, the POCT manager, initiated a lean six sigma project and analyzed the testing workflow. The testing turnaround time was measured, pain points identified, and process user satisfaction surveyed. The solution was to initiate a digital transportation project to provide automated testing and integrate the results with the electronic patient record.

Plan

Project planning commenced (Figure 3.3) involving the POCT laboratory team, hardware and software vendors, and other technical subject matter experts from cybersecurity, infrastructure, integration, and IT operations. A collaborative approach (Skulmoski, 2022) resulted in an approved project plan.

Design

The POCT laboratory and technical teams designed a POCT testing process that emerged from the lean six sigma study. The IT teams captured technical requirements like data backup expectations during the design workshops to support the POCT service.

Microlearning

- What is the DMAIC process?
- What is the difference between ITIL response and resolution time?
- How does backup, recovery, and disaster recovery work?

Build

The approved design allowed the servers, integration and POCT devices to be set up. The POCT technical teams built the system according to the approved schedule and design. The laboratory POCT team was not involved in the build but received status reports.

Test

While the POCT system was being built, the team wrote the test strategy and plan based on ITIL practices. Once the POCT system was built, we followed the approved test strategy and plan and tested the system. Only minor defects were found, and all were rectified. Next, the user acceptance test phase began, and the laboratory team was invited to view the innovation for the first time. The test cases and scripts were executed, and the system performed as designed. The POCT project sponsor did not request any changes and approved the system, formally ending UAT testing. The test report was submitted, and the project went into the final phase before the new POCT system was rolled out to caregivers.

Microlearning

- How does application testing work?
- What is the difference between a test case and a test script?
- What are the contents of an IT test report?

Transition to Production

The transition to production phase includes activities required to handover the product or service of the project to the IT operations team for use by the intended users: POCT specialists and other care-givers. We followed the ITIL framework and collaborated with the release manager to implement the POCT system. The focus of the release manager is to deliver the POCT system into the production environment safely and securely (e.g., risk and quality management). We trained the service desk and end-users. We developed commu-nications and escalation plans for multiple go-live scenarios (e.g., an unsuccessful implementation), established a command center, and implemented other risk management activities. The IT change authori-zation board (CAB) reviewed the POCT change request to go live and approved it.

Microlearning

- What is release management?
- How does the project sponsor use the ADKAR model for organiza-tional change?

Go Live, Stabilize, Closeout, and Use

The organization was notified that POCT was going live. On the morn-ing of the go-live, the team met in the command center (e.g., a laboratory meeting room we booked for a week), gave their readiness confirmation, and the new integrated POCT system went live on time.

The technical teams monitored the system (application, hardware, testing devices, integration, etc.), and the system performed as designed. No service desk incident tickets were raised, and no stabili-zation was required after monitoring the first week of use. The spon-sor agreed to finish the project three weeks early. The project closure procedures were followed, and we celebrated project success three weeks early. Caregivers regularly used the POCT system, and the scal-ing strategy was followed to provide a broader range of POCT devices and services.

Innovation Optimization

The team completed the lean six sigma study by collecting data to compare with the original baseline study. There were improvements in turnaround time (e.g., manual entry of results into the electronic record was eliminated through integration) and increased caregiver satisfaction. The POCT laboratory leadership applied for and achieved the challenging ISO 22870:2006 Point of Care Testing certification.

Microlearning

- Search for Cleveland Clinic Abu Dhabi POCT.

Conclusion

We can expect an increased demand for healthcare innovations because AI-facilitated innovations (followed by quantum computing) are making existing technologies obsolete. Project management is a successful pathway to implement innovation (Figure 3.10). Indeed, we find detailed guidance in the PMBOK® Guide for the other elements of the HIMSS Innovation Pathways model (PMI, 2017b). We increase the probability of innovation success by designing-in quality and risk management into our daily project activities. We also benefit by aligning with the ITIL service management framework, which is the natural way IT departments want to implement, support, and optimize digital innovations. We leverage innovation and develop ideas for further improvements. Requests for additional innovations

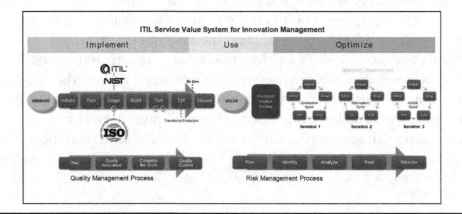

Figure 3.10   Innovation project management.

can be reviewed and prioritized, then taken forward through optimization and ADDIE training sprints. When we use project management to deploy and optimize innovations, we are more likely to be successful.

Notes

1 U.S. Census Bureau. Demographic Turning Points for the United States: Population Projections for 2020 to 2060. www.census.gov/content/dam/Census/library/publications/2020/demo/p25-1144.pdf
2 Fierce Healthcare. Retailers, Payers and Startups Could Capture 30% of Primary Care Market by 2030: Report. www.fiercehealthcare.com/providers/retailers-payers-and-startups-could-capture-30-primary-care-market-2030-report
3 Managing in a VUCA World. n.d. Retrieved August 7, 2022, from www.mindtools.com/pages/article/managing-vuca-world.htm
4 Mintzberg H. The Rise and Fall of Strategic Planning. New York: Macmillan, 1994; 60.
5 LexisNexis Legal Insights. State Legislators Step Up Efforts to Deal with Rising Healthcare Staffing Shortages. www.lexisnexis.com/community/insights/legal/b/thought-leadership/posts/state-legislators-step-up-efforts-to-deal-with-rising-healthcare-staffing-shortages#:~:text=With%20more%20than%20half%20a,the%20next%20five%20years%3B%20and

References

Castellion, George and Markham, Stephen K. (2013). New Product Failure Rates. *Journal of Product Innovation Management*, 30(5): 976–979. https://doi.org/10.1111/j.1540-5885.2012.01009.x

Keeley, Larry, Walters, Helen, Pikkel, Ryan, Quinn, Brian, and Quinn, Brian (2013). *Ten Types of Innovation: The Discipline of Building Breakthroughs*. New York: John Wiley & Sons, Incorporated.

Project Management Institute. (2017a). *Pulse of the Profession® Success Rates Rise: Transforming the High Cost of Low Performance*. Newtown Square, PA: Author. www.businesswire.com/news/home/20170208005615/en/PMI-2017-Pulse-of-the-Profession-Project-Success-Rates-Climb-Fewer-Dollars-Wasted (accessed 30 August 2022).

Project Management Institute. (2017b). *A Guide to the Project Management Body of Knowledge*. 6th ed. Newtown Square, PA: Project Management Institute.

Skulmoski, Gregory J. (2022). *Shields Up: Cybersecurity Project Management*. 1st ed. New York: Business Expert Press.

Skulmoski, Gregory J. and Walker, John (2023). *Cybersecurity Readiness: A Project Management Approach to Training*. 1st ed. New York: Business Expert Press.

Chapter 4

Collaborate and Listen

*Listen for ideas that will potentially solve a problem or present an
opportunity to collaborate with stakeholders and galvanize your network.*

Many innovations started by listening, observing and then communicating
ideas and solutions to problems. When you listen, people are more likely to
share ideas and provide encouragement. The more you engage others, the
more ideas you are likely to catch. Great innovations are typically a result of
multiple iterations by numerous individuals invited to participate in ideation
and execution. Inviting others to share in your innovation will galvanize
support and engagement necessary for success.

One is too small a number for innovation. Innovation is largely a result of a
team of teams' approach to solving a problem or exploiting an opportunity. It
can be an ego challenge to have a great innovation and allow others to modify and edit your dream. We can take innovation too personally and become
captive to the potential and miss out on something greater. Leveraging others
actually frees the innovation to grow and expand beyond what you initially
envisioned. There is strength in seeking the wisdom of others.

* * *

Sepsis Bundle Compliance—Improving Quality
Through Multidisciplinary Collaboration

Charles Zonfa & Michelle Evans

The principles of continuous improvement and innovation have become
the hallmark of an effective quality program. Throughout the healthcare

DOI: 10.4324/9781003372608-4

industry, these concepts have driven transformative change and have reshaped the way in which healthcare professionals view the important and challenging work they do every day. The application of process improvement methodology has become commonplace in the approach to patient safety and quality.

Summa Health, headquartered in Akron, Ohio, is an integrated healthcare delivery system with 8,000 employees, 1,000 credentialed physicians, and a network of ambulatory care facilities providing services to a five-county area. Embedded within the DNA of Summa Health is a desire to establish Summa as an organization of continuous learning. Through the Summa Health Performance Solutions department, numerous physicians, nurses, and staff have completed Green Belt projects to improve quality, ensure safe care, and remove redundancies and waste from the healthcare system. Through this comprehensive approach, Summa Health has been able to translate innovative ideas to improvements that promote better patient care and better health of our community.

The Mission

Sepsis is the overwhelming and life-threatening response to infection and can ultimately lead to tissue damage, organ failure, and death. Approximately 30% of patients diagnosed with sepsis will not survive. Of those individuals who do survive, up to half can have long-standing complications. Considered to be one of the most-costly hospital conditions to treat, sepsis is the number one cost of hospitalization in the United States with sepsis-related costs estimated to be $62 billion annually. Early detection and appropriate management in the initial hours of sepsis have been demonstrated to improve outcomes.

The Surviving Sepsis Campaign (SSC) recommends implementation of four bundle elements as timely as possible, including administration of antibiotics within the first hour of ER triage. To be effective, timely identification of the signs and symptoms of sepsis is crucial.

Activation of treatment bundles immediately after sepsis is recognized has been shown to dramatically improve outcomes. For example, every hour delay in starting antibiotics significantly reduces survival, pushing the mortality rate well above 30%. These treatment bundles have become the gold standard for effectively managing sepsis.

Sepsis and septic shock require specialized hemodynamic monitoring and aggressive treatment necessitating most patients to be managed in an intensive care unit. The hallmarks of treatment are fluid resuscitation, infection control, and antimicrobial therapy. According to the research from the Surviving Sepsis Campaign (SSC), the rate of mortality was 26% in the emergency department and approximately 40% in other areas of the hospital, with rates as high as 44.2% in intensive care units. Consistently, early administration of antibiotics has been shown to have an impact on reducing mortality. Focusing on early detection of sepsis, both in the hospital setting and the emergency department, is essential in any sepsis bundle program. Compliance with initiation of sepsis bundle elements has been utilized to assess the effectiveness of these programs and to gauge program support.

The challenge of any innovative program in the healthcare setting is acquiring enough support throughout the organization to disseminate widespread adoption of best practices. In the case of promoting and adopting sepsis bundle utilization, coordination of multispecialty teams in key departments, including the intensive care unit and the emergency department, is vital. Finding individuals to collaborate across functional hospital departments and to serve as program ambassadors can greatly improve successful dissemination of best practices.

In the pursuit of improving compliance with sepsis bundles, Summa Health embraced the concept of collaboration throughout our organization. We have identified individuals to serve on a multidisciplinary team to tackle this issue and to develop a coordinated approach to the management of sepsis. The success we have experienced in sepsis bundle compliance was dependent on the following: (1) Collaboration between the emergency department and the intensive care unit, (2) promotion of best practices by our leadership champions from both the clinical teams as well as the Summa Health administration, and (3) willingness to listen and modify our process based on feedback from key stakeholders. Our principles of success—*collaborate* and *listen.*

The Structure

The Sepsis Program at Summa Health was formally launched in January of 2021 under the direction of the chief nursing officer and the chief medical officer for quality. Early evaluation of the clinical metrics revealed the need for a more formalized approach to the management of sepsis at Summa Health. As a result of the evaluation process, the need for a dedicated coordinator who could lead the Quality Improvement Initiative for Sepsis Care was identified as an essential element to establishing a successful sepsis program. At this stage of the process, there was a functioning multi-disciplinary Sepsis Committee in place with representation from both the emergency department and the critical care unit. The primary function of the sepsis coordinator was to integrate stakeholder teams by building on previously established work and incorporating innovative actions to propel the continued improvement of sepsis management.

From the onset, it was of extreme importance for the sepsis coordinator to meet with key stakeholders, including Summa Health leadership, emergency department leadership, and individual members of the analytics team from the quality department. Physician and nursing "champions" were identified through this initial phase. Attention was then directed toward the engagement of critical care leadership to assist in examining all aspects of the sepsis care continuum at Summa Health. A collaborative approach was deemed critical to success of any initiative to ensure "buy in" and sustainability of any resulting process.

The Process

Any good process improvement initiative begins and ends with data capture and analysis. A comprehensive and meaningful database is crucial to understanding the process and evaluating tests of change. One of the most important first steps was to understand the magnitude of the problem. The Summa Health sepsis coordinator engaged the quality department data analytics team to develop a process to identify potential sepsis patients (mostly presenting to the emergency department) and establish a daily, real-time log. A database and reporting mechanism was established.

This initial database was utilized by the sepsis coordinator to track and monitor the care of patients with the diagnosis of sepsis. This patient identification list served as the basis for a comprehensive clinical review. First, established exclusion criteria were used to remove any patients from the list that did not fit the standard and accepted CMS definition of sepsis. Second, the electronic clinical record review allowed for determination of sepsis bundle adherence. With this valuable tool, the sepsis coordinator was able to locate patients diagnosed with sepsis and to establish a mechanism to record sepsis bundle passes and failures. More importantly, there was now a record that clearly demonstrated the reasoning behind sepsis bundle adherence failures. This beneficial information could serve as the basis for tracking and trending and could also be utilized as a feedback and educational tool for all healthcare personnel. The next question to answer was how to provide that feedback in a constructive and meaningful manner.

Listening Skills Are Important

The sepsis coordinator created a system to provide notifications to providers (physicians, residents, and advanced practice providers) of bundle adherence fallouts with the goal of notification as well as education. Email notification was utilized to communicate this information and to celebrate successes. Notification usually occurred within 24 to 72 hours of the case. Initially, a few providers interpreted these emails as punitive, regardless of the sepsis team's intent to not place blame. We listened. Based on the feedback regarding the process of notification, we engaged the physician champions to assist in development of a process that could both educate and inform for the purpose of achieving a better patient outcome. The result was a demonstration of support for our provider peers and an opportunity to align goals of the sepsis management program. In addition, it allowed for the opportunity of the physician champions to provide one-to-one education for providers and resident physicians in a non-punitive manner. As a result, providers have a greater understanding and acceptance of the notifications and have begun to rely on them as a method for improvement.

Being a Champion Is Contagious

Shortly into the sepsis initiative, a strong working relationship between the sepsis coordinator and the emergency department sepsis "champion" developed. This relationship allowed for an expedited relaunch of the sepsis protocol in the emergency department within two and a half months of starting the sepsis program. The program relied heavily on proper education of the entire emergency room team and led to the formation of a sepsis alert team. This renewed energy and focus ensured a successful relaunch of the sepsis program as demonstrated by bundle adherence within the emergency department increasing from 44% (as reported by CMS via abstraction through Premier) to the mid-70%. The success of the program when discussed in the sepsis committee piqued the interest of the medical director of the medical intensive care unit. As word of the initial success continued, additional physician champions were identified and engaged.

Identification of energetic, motivated physician champions was one of the vital components to the success of the sepsis program at Summa

Health. We currently have one sepsis physician champion for the emergency department, two champions for the critical care units, and one for the resident physicians and inpatients outside the intensive care unit. These physicians have an established line of communication with the sepsis coordinator allowing for real-time feedback and data to be communicated in addition to present needs, areas of concern, or compliance failures to be addressed. This infrastructure assisted in establishing a method of communication to all providers regarding bundle compliance "passes" or "failures," with the opportunity for intervention and education by the physician champions.

As an added benefit, a method of personal provider coaching with the physician champions was established. The support and involvement of these physician champions has proven invaluable, creating a strong backbone for the program structure and driving sustainable improvement.

Utilizing Results to Drive Improvement

Since beginning the initiative, sepsis bundle adherence has improved from 44% (prior to the Sepsis Program) to 80%, as outlined by the most recent figures reported in the CMS Hospital Compare Sepsis Care data. Additionally, an impact on sepsis mortality at our institution has also been observed. "Present on Admission" sepsis mortality (COVID excluded) has been consistently below the Ohio goal of less than 15%; this number at Summa Health is currently 8.42% heading into the end of third quarter, 2022. "Not Present on Admission" sepsis mortality has also made gradual improvement to around 20% from an initial rate of high 30%–40%. As part of the communication strategy, these goals and results continue to be discussed at departmental meetings throughout Summa Health, along with recommendations to assist in continuous improvement.

Continuing education plays a significant role in improvement of sepsis care and must include a multidisciplinary approach. For example, resident physicians participate in targeted education at the start of their critical care clinical rotation. Moreover, the critical care physician champions have included sepsis recognition and bundle compliance as part of the resident and medical student didactics. As a messaging reinforcement tool, reference cards were created to be given during educational sessions.

This consistent method of education has extended to the emergency department and critical care nursing staff who participate in extensive training on early sepsis detection and bundle implementation.

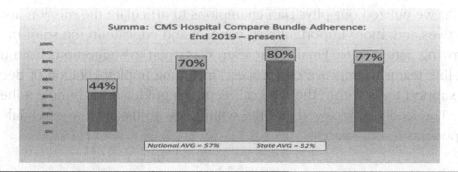

Figure 4.1 CMS Hospital Compare Bundle Adherence

Plan Do Study Act

Led by the Summa Health sepsis coordinator, a Green Belt project was developed to examine continuous improvement strategies. An analysis was completed to identify trends in sepsis bundle adherence failures. A multi-disciplinary team was developed to study potential improvements to the established process. As a result of those discussions, new components to the process, such as instituting checklists, were introduced. Successful implementation led to the expansion of checklists to both the emergency and critical care departments. Nursing education began in the weeks prior to the trial with the goal of establishing a reminder system alerting team members to bundle elements that needed to be completed. The result: Bundle adherence increased to 90% if a checklist was used. While numerous examples of other changes followed with varying degrees of success, the teams are committed to continuous small tests of change to achieve the goal of bundle compliance improvement and better patient outcomes.

The Cycle of Improvement

Since beginning this project, we realized the value and impact that a cross functional collaborative team aligned under a common goal can have. The success of the sepsis project at Summa Health resulted from many distinct components. First, alignment under a coordinator dedicated to both the project and to continuous improvement was the first major step toward success. Second, champions for the cause are essential whether they come from physician leadership, nursing leadership, or the administration. Their ability to set the project priorities and drive effective change is essential. At Summa

Health, we utilized our physician champions to articulate the mission and objectives and, more importantly, to counsel and educate in the spirit of improving patient care. Finally, our team was open to suggestions from the front-line teams to improve our process, including implementation of decision support tools within the medical record to guide sepsis bundle adherence. The feedback received and the willingness to listen were essential components to our success.

* * *

Shadowing to Determine Known and Identify Unknown Issues

By Brittany Partridge

While implementing my first project across a multi-hospital system as a lead clinical informaticist, I learned a vital lesson about workflow and observation, which has profoundly impacted how I work and teach. As I have grown in my career, it is something I attempt to instill in my teams and consistently share in conference talks. The importance of shadowing and workflow analysis as catalysts for innovation cannot be overstated.

The IS analysts and I were sitting in the hospital command center on day three of go-live, riding the high of what we thought to be a very successful implementation of the new ePrescribe system. Very few tickets were coming through, the physicians seemed to be adopting the workflow readily, and we weren't getting many call-backs from pharmacies or patients. I pulled up the daily report of eRx failures to inform where I would go on my rounds to focus on education and troubleshooting and was met with a startling realization. While most of our hospitals had less than a 1% error rate on eRx transmissions, one of our hospitals was running a 30% failure across the interface, impacting the discharge prescriptions of hundreds of patients. This was not good.

I quickly pulled the technical team together to analyze the interface messages; no errors were being thrown. Nothing could be detected between the successes and failures by the naked eye looking at HL7 messages. We continued to try and find a common cause, yet on the failures, the drugs were all different, the physicians were all different, and the pharmacies were all different. I knew then that it was time to head to the units. Technical troubleshooting can only take you so far until user interaction with the system has to be better understood.

I watched multiple providers enter prescriptions, looking to see if there was something unusual with the orders or any shortcuts the users were navigating, but nothing. Finally, nearly at my wit's end, I went back through my flowcharts after a fill of troubleshooting. Who else entered data into the system that affected the downstream prescription process? Analyzing each process box, it hit me: Registration. Off I went to talk to the reg team. After talking through their processes, I still couldn't find a discernable difference, but I noticed there were three registrars, and a light bulb went off in my head. With the error rate being one-third, was it possible that a single registrar was causing the errors? I cross-referenced the errors from the report and found that, sure enough, a single end-user had registered almost every patient with an ePrescribe error. The only challenge was when each user walked me through their workflow; all three seemed identical.

All out of options, I asked if I could hang out and watch the registrar's work. At that point in time, our hospitals were on an old DOS prompt system for registration, which required keyboard entry to navigate the different fields. As I watched the users navigate their screens, I noticed that one was switching through the fields with a different command than the other two. After more discovery, we found that that specific navigation shortcut created a very hard-to-see extra space at the end of the patient's name and address boxes. This space went across the REG-EHR interface just fine but caused the ePrescribe interface to fail. After talking to the user about changing their navigation shortcut to match the other two registrars, the interface errors dropped to align with the other hospitals. I continued to check in with them, as muscle memory is a challenge, and when errors popped up, a simple reminder took care of it. Without going into the registrar's office and watching the different keystrokes, the root cause of the errors would have remained elusive.

It has been instilled in me from my early clinical informatics days how essential workflows are to implementation projects, but I use the ePrescribe example to highlight just how important it is to physically go to the floors and observe, not just talk to the users of your technology solutions. While that example highlights why shadowing and observing are key in implementation and troubleshooting, I wanted to share some other examples and things to remember/lessons learned regarding workflows and shadowing.

Further examples of the importance of shadowing and workflow analysis are as follows:

1. *Current state documentation for pain scale overhaul.* I shadowed each clinical service line to answer two questions: (1) Where do you look for pain documentation in the EHR, and (2) What do you do with this information? During shadowing, I realized just how many places this information was documented and how many "next steps" were completed after reviewing the information. This was vital to help inform a future state design that took into account all the necessary points in the EHR.

2. *Current state documentation for a jail telemedicine project.* While visiting the jail, I noticed that the guards were taking the paper prescription from discharge at the hospital and putting it into a locker (this step hadn't come up in any workflow discussions). When I asked them about it, I found it was for inmates going home a day or two after their hospital visits to get their meds outpatient. This sparked an entire discussion about how this process could be handled virtually.

3. *Trouble-shooting a telemedicine visit.* I had a physician call me very upset that he didn't have the "Join Video Visit" button available during his encounter. I quickly hopped on a zoom call with him to view his screen, and it turns out he had scheduled an in-person visit. So I put him on hold for a moment, called his front desk, and had them send the patient, who was patiently waiting in the physical waiting room, back to the exam room.

4. *Implementing a process to streamline surgical scheduling.* Going to the surgeon's (who had privileges at the hospital but wasn't employed by the hospital) office and observing their schedulers. Observe the hospital schedulers go back and forth trying to get a case on the board, realizing that the information needed and the information asked for in the scheduling forms didn't line up, which created delays.

5. *Implementing telemedicine workflows*: Watching an in-person clinic and realizing how often the front desk would verbally shout back info to the physician, leave a sticky note on their desk, or come over and point something out in the chart, which was eye-opening in the quest to mimic those off-the-cuff relationships and interactions in an entirely virtual space.

Note: As more and more of our clinical technologies and interactions take place out of a traditional hospital setting or in a more hybrid environment with data coming from patients' homes, virtual care hubs, etc., I would argue that shadowing and observing, especially for current state documentation, becomes more imperative to capture all of the micro-interactions and non-standard ways clinicians and patients use the technology.

Key Take Aways and Lessons Learned from My Times Shadowing

1. Embedding—

There doesn't need to be a project or a problem for spending time shadowing and observing your clinical teams to be valuable. When I first joined the CI team, my boss and I brainstormed the best way for me to gain clinical knowledge quickly and understand how clinicians interact with technology. We landed on the idea of embedding me with a residency team. For three months, I rounded as part of the team; I watched the new physicians interact with the tools and paid attention to their clinical questions (and how tech might impact those questions and processes).

Since it was a teaching environment, it was welcome when I asked questions, and the new residents were more than happy to explain further. The value of this time made a lasting impression on me, and I rolled embedding into future large projects I implemented. For example, when I brought the surgeons live on the electronic health record, I worked out of their lounge for many months, allowing them to ask me questions and see where their challenges were. Trust was built through my constantly being around and scrubbing into cases; this allowed us to speak more frankly to overcome barriers quickly and for me to understand the ins and outs of their workflow. My key takeaway from embedding was that the process verbalized is often different than the process performed; shadowing is the best way to get to the bottom of a challenge and build the trust needed to overcome it. Through just being around, I was able to learn about little requests or optimizations that the users wished for and that we could quickly fix but that weren't brought up as they didn't feel they were "big enough." Small, easy wins came from almost every week of rounding.

2. You Are Not There to Judge

Introduce yourself and be clear that you are not there to judge processes and will not be looking for users "doing it wrong." This has been an essential step as I have noticed that when I don't start with level-setting, the users tend to tell me what they think I want to hear, not what is actually happening. Users also try to revert to the process they were "taught" and not what they do daily. I always reiterate that we are just trying to document how the work is currently being done so that a vital piece of information isn't lost when we implement new technology. Often, processes have evolved out of necessity or friction, and identifying the what and why will help move innovation forward.

3. Users Should Always Validate Workflows

Engaging users in validating and expanding current state process maps is key. I usually use the initial workflow diagrams based on my observations since, as talked about before, what is verbally said is being done isn't always what is happening. Then I walk users through them and make additions and notations. All the fun subprocesses, caveats, and exceptions come out. Sometimes I will leave a giant version of the diagram on the unit or clinic for a week with some sticky notes, allowing users to pop in their thoughts as they arise. I always come back to extremely helpful and annotated diagrams. I also encourage the users to suggest parts of the workflow that are currently challenging and any innovations they think might help. These notes very often become the backlog for product innovation and come about organically through validation discussions.

4. Innovation Affects More Than Just the System You Are Updating

Once the users and I have agreed upon the current state, I walk through what the future conditions might look like with a new system. These are my favorite meetings of all time as it allows the clinical staff to see that IS is very focused on understanding their needs. There is also often fantastic brainstorming about how these changes might impact other pieces of technology. We see this a lot in telehealth, where the focus is on changing the audio-visual piece, but things like orders, scheduling, prescriptions, imaging, and all the systems they touch must be included. As the brainstorming continues, we often identify potential failure points and key education opportunities. *As virtual care expands, this point will become increasingly important.*

5. Every Single Member of IS Should Shadow End-Users

As a clinical informaticist, walking onto a unit to shadow became a regular occurrence. That level of engagement and access is not always consistent across teams. On the surgical pick-list optimization project, I learned that many of my developers had never set foot in a hospital, much less stepped into an operating room. We organized a day for them to scrub in with the surgeons and see their applications in real time. The pay-off was twofold: (1) The devs could personally see enhancement opportunities for their app, and (2) a connection was made about their app's impact on end users and patients. That single day generated more excitement and engagement than any other single effort. While the team you are on might determine the frequency of shadowing, I would argue that each IS member should spend a day per quarter free of meetings following their end-users.

* * *

Making an AI That Listens and Collaborates to Improve Mental Health

Alison Darcy

For all you skeptics out there, you are not alone. A quick glance at our app store reviews reveals that many of Woebot's most passionate supporters started out skeptical, never believing that a quirky little yellow robot could help with real problems. We hear this from our health system partners too, that those clinicians who were most skeptical become evangelists after trying out Woebot for themselves, or with a patient.

This is not a good situation for an app developer to be in! Just as well our company's success does not rely on Silicon Valley–style growth targets to be successful. Rather, we believe in good old-fashioned science, a great patient experience, and fidelitous-to-the-tee cognitive behavioral therapy (CBT) delivery. We are innovating on the established field of self-driven CBT delivery, not in spite of it being delivered by an AI, but *because* of it. The ability to meaningfully engage humans in a therapeutic moment through the honest establishment of what CBT creator Dr. Aaron Beck described as *collaborative empiricism* is going to be the breakthrough that we need to improve mental health outcomes across the population.

First, let's dispel some myths, starting with what we mean by AI. While many would paint a dystopian picture of a robot apocalypse, Woebot is more like an AI that your grandmother would make: Warm, practical, and always with your best interests at heart. While Woebot replicates some of the skills that therapists practice, make no mistake, Woebot is no replacement for human therapy. There simply is no replacement for human connection, and human therapists will always be needed. Indeed, this is the biggest red herring that we encounter in popular discourse. The truth is that there are not enough highly trained mental health practitioners, and access—broadly defined to include psychological and attitudinal factors—remains the leading problem in mental health care today.

We're living in the midst of a crisis in mental health that's only getting worse. According to the World Health Organization (WHO), 1 billion people worldwide suffer from a mental disorder, and depression and anxiety cost the global economy $1 trillion every year. During the pandemic we saw a surge in the number of people suffering with mental health problems and unprecedented and widespread adoption of telemedicine in an attempt to meet it. Sadly, we did not solve the problem: Access increased for those individuals who already had it rather than those who are underserved, and we created an army of burnt out, commoditized clinicians.

To reduce the burden of suffering in the population, we must think about mental healthcare differently, make better quality, self-driven tools radically accessible to those who can use them much earlier on in their journey, and ensure that people can get immediate access to human professionals. Finally, we believe that the field of mental health therapy is in its infancy relative to other disciplines within medicine. While access, scalable implementation, and quality gaps are all problems that technology is ideally poised to address, AI is also poised to help us actively improve the treatment models it delivers. In this way, digital therapeutics that use AI are not diluted imitations of what happens in the therapist's office; they are qualitatively different, with unique opportunities like broad reach, convenient and emotional access, consistently fidelitous delivery, and the ability for precision care. If we build these solutions thoughtfully, such that they "listen" and "collaborate" with the people they serve, they could represent one of the greatest public health opportunities we have to shift the needle of the mental health crisis.

Step 1: Choose the Right Therapeutic Approach

Not every therapeutic approach is suited to this kind of innovation. There is only one that has demonstrated wide utility across populations and problems and has a decades-long track record of successful delivery in scalable, self-driven, and recently, digital formats: CBT. Developed by Dr. Aaron Beck in the 1970s, it is based on the idea that it is not events themselves or in isolation that produce mood disorders. Rather, it is how they are internalized such that the treatment helps patients identify the negative automatic thoughts that occur in response to an upsetting situation, see the distortions therein, and replace those thought patterns with more objective, less harmful beliefs. This approach is structured and based on learning skills and problem solving, so it lends itself well to digital delivery.

There's another very good reason why Woebot primarily practices CBT. Some forms of psychotherapy rely on the patient–therapist relationship as a mechanism of treatment itself. In psychodynamic psychotherapy, for example, the patient–therapist relationship will start to replicate many of the elements from the patient's own relationships from early childhood, and this *transference* is analyzed as a key mechanism for change. Systemic family therapy relies on the therapist joining with the family as part of the system of family relationships thereby driving insight through reflecting on the family dynamic. These approaches literally rely on the existence of the therapist–client relationship in the room and therefore are approaches that an AI should never try to replicate. To do so would likely result in a hollow imitation without substantive observation.

For many, psychodynamic psychotherapy is synonymous with therapy itself, but in fact, contemporary approaches like CBT are very different. As with every great breakthrough, it was inspired by talking to humans. Working in the psychiatry department in the University of Pennsylvania, Beck noticed that his patients had a negative bias about themselves, everyone around them, and the future. In the introduction to his 1979 *Cognitive Therapy* manual (Beck, 1979), he illustrates such a bias with one of his patients. If he arrived early for their therapy appointment, his patient would conclude that she was such a mess that she clearly needed more time with him. If he arrived late, she would conclude that he didn't care enough about her to arrive on time, and if he arrived on time, she would conclude that the clinic ran like an inhumane factory. He also rejected the popular notion at the time that people with depression subconsciously did not want to get better and found that actually helping the patient gather the evidence to challenge their negative mindset was often enough to help people recover. Instead of endless analysis of dreams and talking about one's mother, he outlined a therapeutic that was effective, practical, evidence-based, and time limited.

Step 2: Build on the Science That Already Exists

Decades have demonstrated that CBT can be delivered effectively in self-driven formats (including pure "self-help" and "guided self-help") through books, internet, and most recently using app-based technology. Several meta analyses confirm that these can be as effective as therapist-delivered CBT (Cuijpers et al., 2010). However, there are two caveats. First, engagement has been characterized most often in these studies as poor. Second, effect sizes tend to be larger when these interventions include guidance, most typically, from a human therapist who occasionally checks in and redirects people to the correct section of the piece. What this means is that while we have good tools like CBT that can be as effective as in-person therapy, we cannot meaningfully reduce the suffering of mental health problems in the population unless we find a way to make it engaging enough that people are exposed to the correct interactions, whilst delivering the additional therapeutic benefit of guidance in a way that doesn't involve humans.

Step 3: Understand the Problem

To address the significant access challenges, the problem we set out to solve with Woebot was engagement. Not any kind of interaction, but engagement that really matters in the right moment.

When we talked to dozens of people about what they do when they are upset, they told us again and again that they avoid, close the door, disconnect from their families and friends, and retreat into their phones. The engagement we were trying to solve for was *this* moment—how can we invite people to, instead of retreat, practice a skill that can modulate their mood? Our belief is that it is in this moment that engagement—meaningful, wholehearted engagement in a psychological process—matters most. It is this moment that drives the greatest change and is therefore most potent. Many of the digital or self-driven CBT tools that have come before have been based on teaching people things. But learning about how to shift your mindset is not actually shifting your mindset unless you apply it in the moments that produce negative mood states by reinforcing your self-defeating belief systems. Real therapeutic process involves challenging your worldview as all of those negative automatic thoughts are arising. Because being successful in that moment is building the skill to be successful in any moment. Beck called this "State Dependent Learning," the principle that if one learns in a particular state that learning itself is more likely to be generalized to future similar states. So, the more we practice, the more we can literally rewire the connections in the brain.

Of course, we can do a lot of important work in the room with a therapist, but data strongly suggest a strong positive relationship between better outcomes and the amount of work that is done in between sessions (Holdsworth et al., 2014). Prior to the science confirming this, and prior to our recent understanding of neuroplasticity among adults, Beck actually suggests in his 1979 manual that therapists conduct site visits so they can be by the side of their patient when they are confronted with the daily situations in which negative automatic thoughts arise. Most therapists will never do this, but this is the unique opportunity, and privilege, that Woebot has. In this case, traditional exposure-based metrics of how many times one opens an app or how much content they read aren't very relevant. We would rather one moment of real, therapeutic engagement than 10 empty swipe and click exercises. What matters most is the *potency* of the engagement, that is, how wholeheartedly someone can engage and how dramatically they can lift their own mood using these tools in the right moment. Feeling better, then, becomes a leading indicator of symptom change, which in turn occurs before quality-of-life improvement, in concert with earlier stage models of therapeutic change.[1]

Step 4: Understand the Lived Experience

In that moment, a few key things must take place. The individual must recognize that they are in a negative mood state, share that with Woebot (with or without a label), and say why. To meet people where they are, Woebot invites them to share, in their own words, what has led to their negative mood state. In order for the person to meaningfully engage in that moment, the individual must feel heard and understood. They know that they are not judged already—research (Lucas et al., 2014) has shown that this is one of the key advantages of talking to an AI—but they must also believe that Woebot is a credible source of help. Woebot must demonstrate empathy and use an empathic tone that shows a deep understanding of what it's like to be them. To achieve this, our empathy formula is informed by talking to people with the lived experience of the problem that the person is sharing.

For example, in building a program for postpartum depression, we learned that women with postpartum depression experience their symptoms as tremendously isolating and rarely, if ever, talk about it. For those who do, they are often met with a surge of cultural and societal surprise and disbelief; "how *could* you feel overwhelmed and disconnected from your baby?

Being a mother is the most beautiful gift . . ." etc. So, our writing and clinical team ensured that normalization of symptoms became baked into the
encounter. Woebot delivers this normalization, in concert with their character as a gender-neutral robot, using facts and statistics ("I've heard that up to
90%–100% mothers experience intrusive thoughts after giving birth") rather
than more human-style empathic statements ("I feel so sad to hear that,
thanks for sharing that").

Five years on, we continue to collaborate and listen to users to ensure
our products have deep roots in their lived experience. We design for inclusion by starting with ethnographies that study a cross section of people to
help us understand where there may be population differences and to help
us anticipate the potential for harm among marginalized populations. We
listen to users' personal stories to understand something of what it's like to
be them and how they live with and navigate their mental health concerns.

Our current work around sexual orientation and gender identity underscores this approach. For example, we have asked young people what it's
like to come out to family and friends. We've heard stories about getting
rejected by friends or family members, or dealing with people who may be
a little clueless about how to act. By listening to these stories, we understand
what's important to people and the language they use in their lives.

Step 5: Listen and Learn

Woebot was designed to never make assumptions. Where machine learning
is used to interpret natural language, Woebot checks the accuracy of what's
being said, in effect, practicing reflective listening.

For example, if a model identifies that a person has an issue with sleep,
the conversation flow can be designed to help them confirm that endpoint
and reflect. Instead of moving through the conversation, the model can
pause to ask, "It sounds like you're dealing with sleep issues, do I have that
right?" If the answer is no, Woebot can ask for further clarification, and
the conversation can move forward in a more informed manner. Not only
is this how an empathic conversation should go, it also preserves the self-
determination that Ryan and Deci (2017) have shown actually leads to better
outcomes. In my own early research, I discovered that individuals who felt
that they had been involved in the treatment decisions that affected their
care as adolescents were also better able to articulate a vision of recovery for
themselves, compared to those who did not feel their self-determination had
been preserved, even in the context of serious mental illness.[2]

Step 6: Collaborate with the Experts

People are the experts in their own lived experience so, on a 1-to-1 level, that is the ground truth, but AI must listen to and learn from inputs that come from a representative population.

For our postpartum depression product, which received "Breakthrough Device Designation" from the FDA, we very deliberately recruited demographically diverse moms to interview and understand their experiences and to learn and acknowledge where those different issues are. Fifty percent of the participants in our randomized control trial (RCT) with Lucile Packard Children's Hospital were women of color (39% Asian/Pacific Islander; 7% Black/African American; 8% Latina). We're also part of the Scripps PowerMom Consortium aimed at reducing racial and ethnic disparities in pregnancy-related mortality. In fact, in this study design, we are segmenting our recruitment strategy so that when we have sufficient numbers of non-Hispanic white women, we will actually pause enrollment to allow other minoritized groups to enter the study.

We also regularly consult with our Clinical Diversity Advisory Board about methodologies for diversifying recruitment in our studies and what it means to reach a diverse population. The board is composed of clinical leaders with a focus on sexual and gender minorities, people of color, rural populations, and other underserved groups to foster conversation and action that make mental health more accessible.

Finally, we are proud to conduct our own research, both in-house to learn and contribute to the field as well as in collaboration with clinical scientists. To date we have conducted seven controlled trials, have published more than a dozen peer reviewed papers in the scientific literature, and have an ongoing commitment to exploring and developing the role of this technology.

Conclusion

Our mission at Woebot Health is to make mental healthcare radically accessible. Woebot provides immediate support to people through tiny yet transformative conversations that are accessible in dark moments, whenever and wherever they arise. Woebot is a relational agent that has been shown[3] to quickly form a bond with users, similar to that created by human therapists and their patients. Woebot does this not to utilize relationship as a treatment mechanism but to create the conditions necessary to change. When people feel heard, understood, and trusted and when they can see and feel that the tech has been developed by thoughtful humans in the loop, they can take the first step toward feeling better. When we can build tools like this that are effective and engaging at scale, we may actually improve mental health outcomes for all.

* * *

References

Aaron T. Beck. (1979). *Cognitive Therapy of Depression*. 1st Edn. The Guilford Clinical Psychology and Psychopathology Series.

Cuijpers P, Donker T, van Straten A, Li J, & Andersson G. (2010). Is guided self-help as effective as face-to-face psychotherapy for depression and anxiety disorders? A systematic review and meta-analysis of comparative outcome studies. *Psychological Medicine*, 40(12):1943–1957. https://doi.org/10.1017/S0033291710000772

Holdsworth E, Bowen E, Brown S, & Howat D. (2014). Client engagement in psychotherapeutic treatment and associations with client characteristics, therapist characteristics, and treatment factors. *Clinical Psychology Review*, 34(5):428–450. https://doi.org/10.1016/j.cpr.2014.06.004

Lucas GM, Gratch J, King A, & Morency L-P. (2014). It's only a computer: Virtual humans increase willingness to disclose. *Computers in Human Behavior*, 37(2014):94–100.

Ryan Richard M, & Edward L. Deci. (2017). *Self-Determination Theory: Basic Psychological Needs in Motivation, Development, and Wellness*. The Guilford Press.

Hope Lives Here: Advances in Prostate Cancer Treatment

Jon Friedenberg, Jennifer Fox & Edward W. Marx

Prostate cancer is the second most common cancer in American men behind only skin cancer. According to the American Cancer Society, the estimates for 2022 are 268,000 new cases and 34,500 deaths in the United States.

The prevalence means that about one in eight men in the United States will be diagnosed with prostate cancer during his lifetime. But for non-Hispanic black men, the rate is one in six. The disparity in death rates from prostate cancer is even greater with non-Hispanic black men experiencing a death rate that is more than twice that of non-Hispanic white men (37.9 vs 17.8 per 100,00, age adjusted).

Mary Crowley Cancer Research is a not-for-profit, phase 1, solid tumor trial clinic in Dallas, Texas. Phase 1 includes the earliest trials for testing a new drug or agent. As a result, virtually all of our patients are stage 4 cancer patients who have failed or are failing standard therapy. None of our patients receive a placebo; all receive the drug or agent being tested.

It is very common for trials at Mary Crowley to be what are referred to as "First in Human" (FIH) trials. This means that our patients are the first in the United States or perhaps in the world to receive the drug or agent being tested.

To date, there are 20 cancer drugs or agents that have been approved and are now part of standard therapy that began their clinical trial journey at Mary Crowley Cancer Research. In 2003, the announcement that the human genome had been sequenced marked the beginning of a new science called genomics. The promise of genomics is to utilize information in an individual patient's DNA, or the DNA in the cells of their cancer, to formulate a treatment that is more targeted and more effective than standard therapy.

More recently, scientists have begun targeting biomarkers that are not only on a cell's DNA but also on the mRNA or a protein. As a result, a new term, "omics" has emerged to encompass any drug, agent, or therapy that targets a specific biomarker, including one on a cell's DNA.

This new technology means that, at least in theory, doctors and scientists can sequence the cancer cells in a patient and then devise a treatment that targets a mutation or biomarker in those cells. When it works, the only cells impacted are those that contain the specific biomarker that is targeted, leaving all of the other cells in a patient's body unaffected. This is in contrast to a systemic drug that not only effects the patient's cancer cells but also the other cells in a patient, resulting in sometimes devastating side effects.

When this new science emerged, the hope was that one day there would be cancer treatments that were not only much more effective than existing therapies but dramatically less toxic to the patient.

Mary Crowley Cancer Research began conducting omics trials in the mid-2000s. While we continue to conduct trials for traditional, systemic cancer drugs, the percentage of our trials that are focused on a biomarker increases every year. It is estimated that more than 75% of all new cancer drugs or agents under development are targeting a specific biomarker.

In 2021, we opened 47 new trials. This may not sound like a lot, but there may only be 10 cancer centers in the country that opened more phase 1, solid tumor trials than we did that year.

Roughly 90% of our patients are referred to us by their oncologist. The other 10% self-refer.

Each year we open several trials that are appropriate for patients with prostate cancer. Yet in spite of the large number of men with prostate cancer, we struggle to enroll patients in these trials. The main reason is that the overwhelming majority of prostate cancer patients are managed by a urologist and not a medical oncologist.

Our physician leaders are all board-certified medical oncologists with significant experience with early phase clinical trials.

In most communities in the United States, there is little collaboration between urologists and medical oncologists. As a result, it is uncommon for urologists to refer a prostate cancer patient to a cancer trial led by a medical oncologist.

In an effort to enroll more prostate cancer patients in our trials, we reached out to a local urology group that is the leading provider in the region (54 urologists in 20 locations).

We listened to their perspectives on how to provide their patients with more clinical trial options and how to include urologists in our research efforts. In turn, they listened to our approach to conducting early phase clinical trials and our desire to increase the number of trials appropriate for prostate cancer patients.

This led to an agreement to collaborate and establish a new trial clinic focused exclusively on urologic cancers and primarily on prostate cancer. One of the urologists in the group with extensive early phase trials will become the medical director of the new clinic.

This partnership will result in placing more prostate cancer patients in early phase trials. Even with this new clinic, we still faced the prospect of choosing whether we work with a urologist or a medical oncologist on a trial-by-trial basis. For most trials, this is not a problem.

However, there are an increasing number of scientifically compelling trials that require both. For the developers of new drugs or agents that are ready for testing, it is very difficult to find a phase 1 trial clinic that is able to bring together both a urologist and a medical oncologist who will collaborate and co-lead a clinical trial.

Our new strategic relationship with the urology group will, for the first time, enable Mary Crowley Cancer Research to provide the developers of these promising new drugs/agents with the ability to engage both urologists and medical oncologists on the same trial.

It is the hope and expectation of our medical leadership that this collaboration will result in the ability to conduct the most important trials for prostate cancer patients. This in turn will facilitate significant advances in the fight against prostate cancer and ultimately make available prostate cancer treatments that are both more effective and produce fewer side effects.

We believe this new model will not only help patients with prostate cancer but also spark more intra medical specialty collaboration and cure more diseases.

* * *

Career Sauce Secrets: Listening and Mentoring

Helen Figge

I hope this provides an opportunity for pondering some practical thoughts to infuse into your everyday career plans and not only shape your personal career accomplishments but also enlighten others that you encounter in the healthcare professions to do the best that they can do and find their own career destiny. Mentoring holds the key to this success, and the core ingredient in any good mentoring is the art of listening.

The old saying, "If I can do it, you can do it—but better" lays claim to my career path. As a dyslexic hyper-insecure adolescent to an award-winning healthcare executive, it was a journey nothing short of a miracle. With the obstacles I bore throughout my life and career and the challenges that at times overwhelmed me, the only light in a dark career day was someone that I could genuinely reach to for advice and guidance—a mentor. My mentors were of all ages and backgrounds, but all shared one common trait—a special gift to teach, listen and advise. None of my mentors knew me before they became my mentors, but I either gravitated to them or them to me (probably because I was screwing up so badly) and changed the trajectory of my career and life. People don't realize that mentors come from diverse backgrounds with some of my best mentors those that held very low positions on the pecking pole. Yet these individuals saw things no one else could visualize and see in me with personal life experiences that I could learn from and relate. The best mentors are good listeners and taught me the "how and whys" of doing things with an end game of keeping yourself whole, dignified and with a career momentum intact.

In hindsight, we need to realize that throughout the golden years of civilization, "apprenticeships" were established to teach (a.k.a. mentor) those needing to cultivate certain invaluable skills to learn trades, skills and life.

These types of "learning programs" were commonplace in various societies and career tracks, lending way to our civilized development. These apprentices also enabled a continuation of a trade or art that was needed for society to continue to flourish. Given the severe shortage of skilled workers in healthcare and healthcare IT, the success of these "programs" of the past should now be reintroduced and realized as a possible business use case in healthcare in order to have sustainable healthcare models.

Steve Jobs, Founder of Apple, once said, "A lot of times people don't know what they want until you show it to them." Mentors do that by lessons learned and career paths chosen, sprinkling in catastrophes encountered in their own careers. Mentors bring a practical approach to something you otherwise can't comprehend yourself. A real mentor listens but does not preach.

Today, due to staffing shortages and an intense demand for healthcare services, we are accepting more diverse skill sets into the arena of patient care, relying on technology to supplement any lack of healthcare skill sets. The practical experiences that these individuals infuse into healthcare creates optimism, hoping technology can fill in any voids. But the missing piece is the "human factor" (been-there-done-that sort of learning). This infusion of new skills brings with it more of the healthcare industry adopting various technologies to replace a previous skilled labor force to assist but, in the end, not quite meeting the expectations of the market. Enter mentoring. Mentoring enables simple "people factor" skill cultivations for which today's innovative technologies cannot meet the market demands. Mentoring is simple yet powerful and incorporates key concepts such as active listening, building trust with encouragement and, most of all, inspiring a person to reach their career potential.

I am a great example of successful career mentoring. I started out as a pharmacy student in Boston, the "Medical Mecca," lowest person on the clinical pecking order at the time and always surrounded by brilliant individuals like Nobel Laureates wanting to change the world. My survival skills took on a new life form, and I learned the "tricks of the trade." I spoke little and only through listening, watching, observing and doing many tasks that no one wanted to do was I able to survive and excel. I always struggled with school, having to take extra hours for studies that otherwise took someone much less time; I always needed visuals and could never see or think abstract. I suffered from dyslexia as a child, and I always had at least two jobs to pay my bills just to survive in Boston as a college student. It was very intimidating at times because the "brain power" around me made me feel even more inferior.

Often, I was too embarrassed to ask a question. But one thing was in my favor, despite my personal barriers—I was a great listener, and people who pontificated a lot loved that about me. I gained the "like" factor quite easily, never questioning or creating a commotion. I just listened and absorbed like a sponge. I recall that I always nodded my head a lot during my early career days, listening intently to what was being said and acknowledging thoughts in conversation. One day on clinical rounds, Dr. Eugene Braunwald (author of *The Heart*, cardiology bible) told me to stop bobbing my head up and down because I looked like a turtle sticking his head out of a shell (that was actually good mentoring, so, note to self, "stop looking stupid on clinical rounds"). I listened and learned. I also volunteered a lot to do things others found too demeaning, but those "volunteer jobs" were my ticket out of mediocrity and getting "noticed" despite my many flaws. The individuals I volunteered my time for were indirectly my mentors because I learned, listened and appreciated what was being said. They also used to say, "listen and you will learn." (I always heard that phrase from my grandpa growing up.) So, through these efforts, I accomplished things others failed at just by sheer hard work and determination. Odd volunteering jobs like compiling data or cleaning test tubes for a Nobel Laureate created a networking and mentoring path that today made my career. Each job taught me something, and that was important to me. These odd jobs I tackled also at times taught me what NOT to do too, how not to treat someone, how not to do something, etc. I perfected the art of listening intently and gravitating to those that I could learn from. To this day, I appreciate those feelings and, in turn, practice mentoring by "listening" most of all to the future generation of technology leaders.

Make no mistake, listening is an art form. So, I learned quickly to speak only if the words added to the current conversation. Listening can be used both ways to be mentored or mentor someone.

I learned several great mentoring tips that I accumulated through my mentors that helped to create the career that I enjoy today. Here are most favorite to share: First, if I was to be at work or someplace at 7:00 A.M., I showed up at 6:30 A.M. and helped the prior team end their shift a little quicker and less hassled. If a deadline was for a Friday, I presented the material on a Thursday. I spoke only when I presented the deliverables and only questioned or got redirects and never asked for extensions. I built up the "like factor" fast with this effort. I also learned quickly that if people genuinely "liked" you, they gave you a break when you really screwed up— which I did often. I always felt the need to help others and always yearned for people to respect and like me. If someone didn't like me or my work, I would go through excruciating lengths to figure out why. I finally gave up on this line of thinking over time as I felt it was not my issue just their issue and moved on.

Also, continued learning is key to good mentoring of others. I learned to self-educate myself with skills that I needed in order to fulfill a job requirement and to be competitive in the workplace. I accumulated most of my degrees on weekends and nights in hotel rooms (if traveling for work) or when I put my kids to bed and was then able to grab a two-hour block to learn something that would help me be better and more confident in the workplace.

If you are a hiring manager and reading this essay, please note that you will never find the perfect candidate for any position; but you can find an individual that has skill sets that can be cultivated into a role that fits your desired qualifications for a position you seek to fill. I found my best hires were the "under dogs," those candidates that might not have been my first choice but because I provided them the opportunity to apply their strengths while being "mentored," proved beyond a doubt that their drive to excel when given the opportunity surpassed all expectations as if they always were destined to fill the role.

So, you ask, how do you get to this point in a person's career to realize the obvious might not be the obvious? One word: "Listen." Listening comes in many forms not just with our ears but listening to our surroundings. Listen to what are people asking. What are people questioning? Who are they speaking about and why? Is there someone in your environment that wants to fit in but can't because of the environment with which you breed? Listen to what is not being said as much as what is being said. Listen to the noise, not the static. Listen to the audience "ask," then decide if their "ask" is something you need to address, or did they not listen to you because of your messaging? Most importantly, *never* listen to or abide by gossip or innuendos.

Form your own opinions about people, places and things, not just the "hear say" or stories of a person and what someone else's perceptions are based on another's assessment. Form your own opinions by listening to the pure data and circumstances, not how it is perceived or delivered to you by others. Remember, often the guiltiest party in any circumstance is the first one to get the story out. So, remember there are two sides to every story, and a true leader makes an assessment by hearing and listening first-hand.

A final important level-setter is to listen to everyone regardless of title, age or position. When being a mentor to someone or being a mentee, don't worry where anyone is on the "totem pole" of titles, age or position. We can all learn from each other. I learned my best PowerPoint skills from a secretary who did graphic ceramic artwork on her free time; I learned how to book travel from an engineer who developed an app to find the cheap rates because he was extremely frugal; I learned how to run a tight P&L from a communications director who had a limited budget; and I learned how to be a decent individual from my many bosses that I chose to work for, regardless of pay or title. I always chose a job because of what it did for others, not how much I could make in the role. I learned to make my own coffee and carry it in a large travel mug in my trunk while on the road, versus buying beverages on the road; I traveled light and always had a bag in my car of emergency essentials—like granola bars and a bottle of water. All tricks learned from others on the job, indirect mentoring at its best.

For healthcare to be sustainable, we all need to realize that there are different approaches to various problems and some more orthodox than others. There is no easy "fix" but rather a compilation of several different pieces needing to work together to create a unified and streamlined approach to its delivery. With many organizations coming into healthcare, we need to "listen" to the market and the various stakeholders to figure out what is working and what is not working. We are seeing the healthcare world becoming boundaryless yet ripe with novel ideas previously unheard of at the time. The art of "listening" will be paramount as we reshape healthcare, and its destiny. Listening is the core of all great mentoring.

So how do you "learn" to be a good listener? Believe it or not, it is a challenge for many today. First, always set aside time to meet and greet people, places and things. You are never too busy to provide your time to others because if you love what you do, time never stands still. Just stand quiet for a few seconds and survey your environment, whether it be in a meeting, in line for food or in the board room, and just notice people places and things; feel the energy around you and the inhabitants of the room because energy and vibes play a big part of the "feelings" you will encounter when listening; and seek out those sitting in corners and engage them; ask others their opinions; confidently look everyone in the eye when speaking; remove all disturbances like social media technologies; and most importantly, thank everyone for their opinions. A good listener is genuine and honest and nonjudgmental. Be that leader who is a good listener. It made me who I am today against all the odds I never would get here.

Finally, mentoring doesn't have to be a formal program. It can be in our every day and in every possible encounter we have in our careers. Technology cannot replace the "people factor" that we need for healthcare's sustainability. No number of innovative approaches can replace looking someone in the eye for discussion. The technology we use today enables us to handle data, but it does not enable us to truly deliver the emotional or "people side" of healthcare.

My final thought about career paths and the "people skills" needed to sustain a great career path is about a simple but "golden test" that I routinely use in my career and life. It's called the "supermarket" test, and it goes like this: Say you are walking down an aisle in a supermarket, and you spot someone down the end of that aisle, and you hesitate to continue your path down the aisle because of an experience or past event you had with them. If you can't walk down the aisle because you feel a strong need to avoid them, you have done something wrong. In your ability to move forward in your career destiny, figure out what the issue was and learn from it. I always say to my friends, colleagues, students and family, let the hate and inability to look someone in the eye be on the other side of the fence, not on your side where you live.

* * *

Disruptive Innovation in Healthcare

Peter Tippett

The process of innovation is most simply defined as the "introduction of something new." We often think of the invention of the telephone, personal computer, internet or other monumental achievements as the embodiment of innovation, but the reality is that innovation happens every day at varying scales and with varying impact.

Personal, Local and Market-Scale Innovation

On a personal level, many people, especially those of us who like to "tinker," regularly innovate while solving problems for ourselves, our family or others facing a challenge, with a novel twist on something old or by leveraging an existing capability. We might substitute one mechanical part for another in order to create or repair something or use common software to create a purpose-built solution. In many cases, this might only yield "good enough" efforts for a single or limited use, but they are innovations nonetheless.

At a company, local government or on another limited scale, followers of a larger innovation may become the first early adopters, leveraging newly established technologies or capabilities as-designed and contemplated by the technology. Some recent historical examples include adding online purchasing where it was not previously available or moving manufacturing or customer support offshore to deliver a competitive advantage. Extending newer technologies to adjacencies and/or use cases that they were not designed for is the other most likely type of innovation at the local or medium scale.

The innovations we all know and love are those that happen at market-scale—in which an entire industry, economy or even the world is transformed because of wide-scale adoption. In the past few years, companies such as Apple, Amazon and Google have delivered this scale of innovation: Creating the platforms and tools that are used by personal and local innovators so that they in turn may leverage the operating systems, web or online platforms to deliver on the next big idea. In healthcare, we have long expected that precision medicine, as well as the shift to electronic health records (EHRs) and making meaningful use of health information technology, would also be disruptive. But neither has yielded disruption yet.

How does market-scale innovation happen? First, let's sub-divide large-scale innovation into the following categories:

- **Driving** innovation is how established companies continue to evolve, grow and drive new business, typically into adjacent markets. One example is in the next generation of telecommunications, as with the deployment of 4G LTE as a major advancement in mobile digital bandwidth and capability.
- **Sculpting** innovation is how people or organizations use an existing strength to move into a newly established space in which pricing and other business models are becoming understood and accepted and where the disruptive flux is settling down. An example includes Google, IBM and Microsoft moving into cloud computing after Amazon AWS and others established the concept and market.
- **Disruptive** innovation changes the entire way in which we do things and also typically lowers entry barriers and price while changing delivery channels, market access and player relationships. Examples

in healthcare include the century-old disruption brought by evidence-based care, antibiotics and the pharmaceutical revolution, sterile procedure and the surgical revolution, etc.

For the purpose of this chapter, let's focus on market-scale, *disruptive* innovation in the healthcare industry, keeping in mind that true disruptors rarely derive from a single person but require the skills and culture of an entire organization to achieve market transformation. In my career(s), I have found that beyond inspired leadership, there are common attributes to organizations that are successful at delivering market-leading innovation, including the following:

- A passionate and deeply ingrained vision
- Hiring for strength and diversity
- A culture of empowerment for all
- Both confidence AND doubt (with rewards for open-mindedness)
- Hyper-near-term delivery focus (while never losing sight of the long-game)
- A fanatical focus on and understanding of the customer
- Creating multi-dimensional customer delight
- Unrelenting drive for simplicity
- Willingness (and eagerness) to adapt as the market unfolds
- Solving the surround

Each of these deserves exploration and analysis, but of all of the traits needed to succeed, one of the least well understood and most powerfully real challenges in building a disruptor is in "solving the surround," or understanding and addressing what I like to call the "seven big problems."

Solving the Surround

To explain the concept of "solving the surround," let's consider the evolution of the music industry. The first disruption occurred somewhere in the Middle Ages when concert and opera halls combined with the development of the symphony, powered by new styles of city living. The concert hall model provided scale, reliable monetization and delivery mechanisms, along with a driver for new musical content. Artists and musicians were paid to create and perform, and individuals had a reason and a simple way to consume.

The next disruption of the music industry occurred with the invention of the phonograph. This was combined with radio, which provided a second channel of delivery, as well as new opportunities for monetization via advertising revenue. Now music was delivered *to* people.

Subsequent music innovations, such as LP, Hi-Fi, Stereo, multi-track tape players and CDs were all examples of sculpting innovation. Each of them improved quality and extended the market into adjacent markets, but none of them changed the fundamental model.

The most recent disruption in music was achieved through the success of Apple's MP3 player, the iPod. The iPod did not win the market and drive disruption of the entire industry because it was from Apple or because of its great design, but because the company had also "solved the surround"—to identify and address what would be required for mass adoption and to support new delivery, distribution, monetization and consumption models.

Other well-funded companies, including Microsoft, Creative Labs and Philips, had MP3 players and excellent engineering, marketing and distribution but never realized the same success. Apple had these capabilities but also identified and attacked the "surround" problems, such as the following:

1. Software needed to be available for the PC and Mac.
2. Users needed easy ways to use ("rip") their old CD collection.
3. An online store needed to be created.
4. A huge library needed to be created to house a wide selection of music without needing to purchase an entire album to acquire 1–2 favorite songs.
5. Several different licensing issues needed to be addressed.
6. There needed to be a way to synchronize and backup music across devices, and more.

In other words, Apple broke down most of these barriers at once as part of their "product" to enable broad adoption of their device and platform and deliver rapid revenue during a period of continued innovation and growth for the company. By taking this approach, Apple was the one to successfully disrupt the music industry.

But, disruption cannot happen in a vacuum. The market needs to be "ready" for disruption to be successful so that the old way of doing things is replaced by something new or is supplemented by a new model that is much larger than the old. In music, the success of the concert hall model didn't come solely from the building of symphony halls. It only worked because the new model of urban living was taking hold. And the iPod needed the internet, online payment systems, ubiquitous home computing and much more. The success of the iPod wasn't the result of Apple build-ing all of the things required for the market to be "ready"; instead, Apple identified the few additional "surround problems" that could be leveraged and accelerated for the otherwise-ready market to complete the puzzle and enable the disruptive market shift.

The Revolution That Missed Healthcare

To understand market readiness, let's step back and consider the path that healthcare has taken thus far. In many respects, healthcare IT never had its "PC revolution." We all use personal computers, but in healthcare enter-prises, there has not been much individually driven business use to solve our common healthcare delivery problems. Consider what "personal" meant for the adoption of PCs in most other sectors: When the PC was born, and long before the Internet, the CFO and the finance staff in many large orga-nizations had projects waiting in the IT queue. Perhaps one was related to understanding different pricing models, another might have been related to planning the budget for the next year or maybe something around new expansion or an acquisition. Whatever they were, the finance staff was beholden to the IT staff to get the project planned, scoped, vetted, priced and staffed, then tested, trained and deployed.

Then came the PC. The CFO or another finance leader went out and bought one. After a few days of trial and error, he/she realized that with the PC and some personal software and a few thousand dollars, they could do their own analysis of pricing models, budgets or expansion or acquisition analysis without any help from the IT staff. They simply bought PCs and soft-ware for their team and told IT to spend their energies on bigger problems. And the same scenario played out in law firms, engineering companies and other innovative groups seeking to improve how they got their jobs done. The spreadsheet and word processor were the "killer apps" of the PC era.

Together, the PC, internet and mobile revolutions led to the biggest expansion of workforce productivity since WWII. Productivity in *nearly* all industries soared. The biggest exception was the healthcare sector, which did not participate in that productivity revolution—or at least did not real-ize anywhere near the same benefits as other sectors like finance or retail. Whereas the constant-dollar prices for product and services in most sectors leveled off or declined, the cost of healthcare continued its inexorable rise, and health IT largely kept to the IT-centric model; EMRs and related soft-ware were planned centrally and driven out as a formal IT project.

Bringing Disruption to Health IT

In the 1990s, some doctors and other healthcare professionals began using PCs and dabbled in using them to help with clinical work. They were often worried about the privacy and security of these new technologies and especially about using email to share dictations and other medical patient information. At the same time, electronic communication of medical billing information within the payer sphere was beginning to evolve.

William (Bill) Braithwaite, MD, PhD, then Senior Advisor on Health Information Policy at the U.S. Department of Health and Human Services (HHS), recognized these problems and authored the Administrative Simplification Subtitle of the Health Insurance Portability and Accountability Act of 1996 (HIPAA). He was not aiming to restrict data sharing but to help enable it (remember the "P" stands for portability) by showing how to address the privacy and security problems. I was fortunate to sit on the President's Information Technology Advisory Committee (PITAC) with Bill, which closely coincided with the creation of the Office of the National Coordinator (ONC) at HHS. As part of that group's mission, I became invested in helping the healthcare industry achieve the "Triple Aim": Improving the patient experience of care (including quality and satisfaction), improving the health of populations, and reducing the per capita cost of healthcare.

Most of the PITAC members and I (and many, many others) thought that the physician and hospital markets "were ready" for both widespread adoption of EHRs and for widespread digital communication of both medical messages and electronic versions of medical records. We anticipated the future that EHRs would enable, and that doctors and everyone else would embrace EHRs just like the business and academic world had adopted word processors and spreadsheets. And we believed that records would be shared between and among care teams in larger health systems and smaller practices and include "ambulatory islands" like physical therapy groups, ophthalmologists and skilled nursing facilities. It appeared the adoption would mirror the way in which email and text were used by professionals in other industries.

Of course, HIPAA and PITAC begat ARRA, HITECH, Meaningful Use, MIPS/MACRA and more, which strongly incentivized the deployment of EHRs and the sharing of digital medical records. The drive for broad deployment of EHRs was effective, but the widespread sharing of digital medical records among caregivers in different organizations largely has not come to pass.

"Fast Forward" 20 Years

In late 2017, I had lunch with David Blumenthal, MD, former National Coordinator for Health Information Technology and current President of the Commonwealth Fund. We spent much of the time reflecting and discussing why modern digital communication and medical record sharing is still so abysmal in healthcare. He acknowledged lots of inhibitors but focused primarily on the competitive business issues between health systems and also among health IT vendors.

After that lunch, I was inspired to write him a lengthy email with a list of potential inhibitors and later tried to distill our shared beliefs into a blog post that provides insight into some of the "surround problems" restraining healthcare data sharing:

> *Most people working in the healthcare system are interested in tools that help in getting their job done and in providing quality care. Therefore, most people relish easier digital medical communication. Most scientists, policy makers and regular citizens are Connectors [those seeking to ease data sharing also include independent physician practices, specialists, clinics and probably smaller hospitals]. And with the shift to quality and accountable care, along with incentives from payers and evolving regulations, even some of the Resistors [those organizations who fear the sharing of patient records conflicts with their broader business interests] are getting on the Connector band wagon. So why is the sharing of medical records and messages still so difficult? I see several primary drivers for this market "conflict":*

1. **Market Competition**—One of the dominant early reasons for the lack of easy health data communications has been business resistance by large providers and health IT companies as a competitive maneuver. By seamlessly sharing digital patient records across town with competing facilities, the hospital potentially dis-incentivizes a patient from remaining within their own system. Similarly, the EMR vendor, seeking to become the dominant system in a region, benefits from volumes of patient data staying within their domain to prompt market demand and expansion to neighboring facilities.

2. **The Ubiquity Problem**—The value of any network to its users is a function of the number of users cubed. We wouldn't find phones or

email addresses to be particularly useful if we couldn't reach the right people because a communications network that does not include nearly all appropriate users isn't likely to succeed.

EMR adoption has generated exactly this type communication challenge because as an industry, we have created fairly closed user groups for each enterprise, when one of the largest values to clinicians, to patients and to the market is in the secure flow of medical information where it is needed. Where we are now is like having computers before the internet. They were useful, but adding ubiquitous access and sharing brought by the internet (with proper privacy and security) led to multiple orders of magnitude of expansion, along with numerous extensions of entirely new value.

3. **High Costs**—Healthcare is a business, and across the industry, providers invest heavily each year in modernizing systems, equipment, personnel training, and related operational areas. Since 2009, technology costs at physician-owned multi-specialty practices have increased by more than 40%. . . . After investing so heavily in IT requirements, the industry is understandably fatigued. . . . Until the industry creates truly turn-key secure communications capabilities, resistance will remain.

4. **Enterprise Focus**—In healthcare, the dominant adoption models and drivers over the past 20 years have been focused on business-to-business (B2B) value propositions, sales strategies and implementation methods rather than personal value needs. . . . Broad cultural shifts rarely take place without the buy-in of individuals en masse, and to secure the future of communication and the flow of digital health information, the individual will need to play an active role in market adoption so the technology itself needs to be tailored to the individual.

And there are many more surround problems, such as the challenges of arriving at consistent data formats and labeling, managing hundreds of thousands of unique data elements, the complex relationships between the numerous data elements and the need for contracts and Business Associate Agreements to share information. Clinically speaking, the patient's story or narrative is at least as important as anything in a database. In order to change behavior or create a viable treatment plan, doctors need to understand the "nuance" and the "why" at least as much as the "what." And there is the elephant in the room: The need for security, privacy and compliance to be both effective and universal to protect patient health information.

So here we are, decades past the PC revolution, with a combination of industry standards, regulations, clinician and consumer demands and even tens of billions in EHR incentives, but not yet able to truly disrupt healthcare IT and communications. We have neither a 'killer app' nor ubiquitous medical communications, and as a result we don't have the efficiency nor ease-of-use benefits from our EHRs, nor do we have repeatable examples of improved quality or lower errors.

I am confident that we don't have a market readiness problem. We have more than ample electricity, distributed computing platforms, ubiquitous broadband communications and both consumer and clinician demand. We have strong security and the legal, privacy, compliance, data format, interoperability and related standards to move forward. So I would contend that our biggest innovation inhibitor is our collective misunderstanding about how to solve the surround. Once we do that, we will unleash market disruption and find success.

* * *

Bedside Nurses Use Their Experiences in Patient Care to Drive IT Solutions

Stefanie Shimko-Lin

> *We are useful supernumeraries in the battle, simply stage accessories in the drama, playing minor, but essential, parts at the exits and entrances, or picking up, here and there, a strutter, who may have tripped upon the stage.*

I am a registered nurse (RN) and clinical analyst at the Cleveland Clinic, one of four RNs and a flight paramedic who work in this role in the digital domain of information technology. Before becoming a clinical analyst, I became a bedside nurse. I've been a bedside nurse for ten years of my career and have witnessed births, held the hands of the dying and all that is in between. Every patient I've cared for has touched my heart deeply, and I have a story for all of them. Not all nurses have a chance to work in IT, and not all IT personnel get to step into the shoes of a nurse. I'd like to share a patient story with you for which I remember every detail. It led me to the William Osler quote at the start of this essay and has shaped my career ever since.

I was a night shift nurse on a cardiac step-down unit, mostly postoperative or preoperative open-heart patients. I vividly remember a woman who passed away in our unit two days before Christmas. It was the first time watching a life slip away in front of me after the valiant fight to save it. My personal goal as a nurse was to know the person I was caring for before the end of my shift. I knew this particular woman's family, how they would spend Christmas and the hopes that her surgery would give her the energy to enjoy her life, children and grandchildren. She was in her late 60s with happy years to live, or so I'd thought at the time. I'd met some of her family members and knew a bit about her as a person. We were unable to save her.

One moment she was using the bathroom and the nurse was walking her back to her bed when the patient reached her hand out in front of her and went limp. An alarm sounded, and we all ran in to help. We got the patient into her bed and began CPR and called a code in an effort to restart her heart. A cardiology fellow and friend of mine ran the code that day. We tried our hardest for 40 minutes to revive her but were unable. The feeling is one of a deep pit in your stomach. Doing all you can, only to fail to bring back a life that existed only moments before. We readied her to have her family come and say goodbye to a mother they hoped would be there for Christmas, for weddings, for births—now gone. We cleaned up the patient and combed her hair, changed her gown and sheets. It was surreal watching the physicians call and ask the family to come to the hospital, watching the family arrive thinking their mom was still alive and seeing the face of the physician with tears as she walked down the hall to break the news that their family member had died.

The physician breaking the news was a resident, and she said, "Oncology patients are easier to do this with. At least then you know it is coming and it's not a surprise." I'll never forget that statement as my mother died from stage 4 uterine cancer a few years later, and I believe it is a little easier knowing, but not by much. I remember the scene as if it were yesterday, though it happened almost ten years ago. The fellow came out to the nurses' station, and we were talking about what occurred during the code. He had been remarkably calm and never raised his voice once. Anyone having been in a code might tell you that is not always the case. It's stressful, lives are literally at stake, and it can be chaotic at times, but not that night and not that physician. I asked why he was able to remain so calm, and he said he'd graduated from Johns Hopkins, and part of their training was to remain calm in the storm: Aequanimitas. He had a pin he wore on his lab coat that had the word on it. He told me to read William Osler's essay "Aequanimitas." I urge you to read it if you never have. The earlier quote is one I have taken to heart and feel I have completed my role as a nurse and then some and know my purpose.

After reading the quote by Osler, I realized there was a correlation between my patient in the scenario I described and Osler's quote. My patient's life was the play, and I was merely playing my part in her story. Many other families have had similar stories, and other nurses and physicians played their parts in those stories. As direct care providers in a clinical setting, you pay attention to the details no matter how minute and develop an urgency for how the storyline will play out. We are trained to take responsibility for others' lives in healthcare, and it can be a heavy weight to bear at times. But with that responsibility comes the ability to see clearly what's important and what to prioritize. One small change in the storyline, and the whole play changes.

If we are stage accessories in life's play, then the medical record is the script for each player's life. As a clinical analyst working in information technology, it is of utmost importance to me and to other clinical analysts on my team to make sure the script is written in the best and most efficient way and that the information is used to provide the best care possible for our providers and patients. Our team of analysts and developers have created tools that help tell the patient's health story and surface information, which was already recorded in medical records to enhance outcomes and patient safety, as well as provide comfort and meet patient requests, even at the end of life.

Vital Scout[SM] is one such tool, created to help identify patients whose health is subtly deteriorating and surface the information in a way that nurses and providers are able to clearly see there is a problem and intervene as soon as this is noted. Vital Scout[SM] was the brainchild of one of our nurse clinical analysts. Her background is critical care and pediatrics. The patient stories she knows and has heard led her to do research on early warning systems and the information they use to identify patient deterioration. As part of her master's degree dissertation, the early warning system was used, and she felt it was an important part of the patient's chart. It would help nurses and physicians intervene sooner for our patients and identify problems early on. Her passion for patient safety and nursing drove her to create a risk-stratification tool based on five pieces of patient care information discretely documented within the EMR. Vital Scout^{SM} provides users on medical-surgical floors with a visual cue for subtle changes in a patient's vital signs and correlates the changes with a stoplight color scheme. Users can escalate a patient when the patient is doing poorly. Interventions are performed based on which criteria are considered out of normal limits, then assessments are required within a specified time frame. A report within the EMR is used and seen by nursing as well as physicians, and providers are able to see what vital signs are changing and causing a score change. All patient care providers are now on the same page when caring for patients.

IT included nurses from pilot units in building the user interface for the tool. The lead clinical analyst met with users in a medical-surgical unit to get their feedback on what would be useful to them in reacting to a patient who is decompensating. Feedback was obtained by finalizing a pilot version of Vital Scout^{SM} used in one of our regional hospitals. The pilot confirmed the ability for an early warning risk-stratification tool to decrease admissions to ICUs from medical-surgical units. The tool includes a screensaver that shows a visual representation of patient scores on each unit without jeopardizing patient privacy. It provides situational awareness and helps nursing become a team sport. Now, nurses on every unit have eyes on their patients. From those in environmental services to physicians, nurses and anyone seeing the screen saver, they can identify patients with changes and speak up so that problems can be corrected before a patient declines and requires more invasive treatment and in a worse case, death.

Vital Scout^{SM} ended up being a particularly important part of the caregiver/creator's life when she found herself in the hospital as a patient. She had an infection, and caregivers reacted when her Vital Scout^{SM} score turned to blue after a blood pressure had been recorded low and out of the norm for her. Immediately her nurse received orders from a physician and started IV fluids to correct the problem, and her BP was restored to normal limits after the intervention. Once again, a project generated by a nurse in collaboration with nursing informatics and bedside nurses improved the outcomes of a patient and, in this instance, not only a patient but the patient who created the tool. The early warning system exists because of the passion generated by the experiences of our nurse analyst.

Iris Behavioral Health was a project created in collaboration with Nursing Informatics and bedside caregivers. At the time it was created, all behavioral health units had their safety round documentation performed on paper. Real-time computer documentation was needed. A previously created application called Iris was used in a past documentation pilot with the iPhone. Nursing Informatics was familiar with Iris and requested assistance with creating an application on a mobile device. Specifications for the information needed were provided to our team. The team building the app consisted of two RN/clinical analysts, a systems analyst and a software developer. Because we were familiar with nursing workflows, we were able to make suggestions to the systems analyst and developer and create a unique application that improved the safety of both care providers and patients alike. Iterative development between Nursing Informatics, our team and bedside staff took place, and the final solution was created. It is currently used in all behavioral health units in our hospital system.

Nurses have a lot of tasks to do during any given shift. The final application included a color-coded timer indicating how long it has been since the last safety round and served as a reminder for the nurse/care provider to check on the patients every 15 minutes. One of the nurses working on our pilot units made suggestions based on how they do things on their unit. Thanks to this floor nurse, the application was approved to have one question, requiring three taps on a phone screen per patient, one to open the patient, one to select the patient's location on the unit and one to save. During training, buy-in from staff members was increased when we introduced ourselves as active nurses. As nurses, we know how new workflows are received by staff, and we were able to incorporate those concerns into our training. The outcome was a successfully used application that increased patient safety. During a staff meeting in early 2018, one of our nurse leaders relayed a story about a patient who was admitted to one of our behavioral health units. This patient's life was saved when a staff member performing safety rounds entered a patient's room, and the patient was actively attempting to take their own life. The caregiver entered the room because according to their device timer, rounds were due. Currently, the application has increased patient rounding from every 15 minutes to an average of every 10–12 minutes. The decreased time between safety rounds limits the opportunity for patients to perform self-harming behaviors. The success of this application is a direct result of different parts of nursing and information technologists working together to make one solution.

The EOL (end of life) project in particular is very near and dear to my heart. Our team was contacted by the organ donor coordinator for the

Cleveland Clinic. She educated us on the need to call organ procurement with each patient expiration in the hospital setting and how organ donation affects lives. One donor has the ability to save eight lives and affect up to 50+ more with tissue donation. She asked that we help surface basic information already documented in the EMR by nursing and physicians in order to better serve dying patients and their families at the end of life. In addition to patient concerns, CMS requires that organ procurement agencies be called after all hospital deaths. Without a way of pulling information into a report or filter, there was no way other than manually searching each patient's chart for information regarding their status and making the call after the fact.

Three documented pieces of information are used to identify potential donors: Two missing brain stem functions, the patient is on a ventilator and the patient has a life-limiting diagnosis. Each referral only allows the family liaisons to discuss organ donation with the family and provide information for informed decisions as well as find out if a patient wanted to be an organ donor. It also allows Lifebanc to review a patient chart and see if they are a potential organ donation candidate based on their medical history. As a former hospice nurse, I have a great respect for end-of-life care and giving each patient the dignity and considerations to make their final wishes a reality. Organ donation is a last request of sorts, and if we fail to meet that request, then we fail the patient. This is another example of how the passion for patient advocacy sets the stage in IT. The tool we created efficiently surfaces information considered when making a candidate known to Lifebanc and our coordinator/family liaison team.

Call information can be seen by all care providers, allowing us to make calls sooner and give families more time to consider donation as an option if the patient did not explicitly state their intent to be an organ donor. Currently, if a patient wants to be a donor, the team at the hospital only has an hour to contact Lifebanc for a referral and only three hours once a patient is pronounced brain dead. Time is of the essence when these patients are identified, and our workflow is important to meet the time constraints of the process. In another situation of life imitating art—in this case the artful care of a dying loved one—I had the honor of witnessing the referral process for organ donors first-hand when a family friend passed away suddenly a few months before writing this essay. The family had some closure and comfort knowing their daughter helped save seven people with her organs and will help many more with her tissue. My experience with the patient's family also shaped my approach to the project.

Sir William Osler had this to say about nurses, and I believe it applies to those of us in healthcare who put our heart and soul into caring for others whether at the bedside or behind the scenes in IT:

> *There should be for each of you a busy, useful, and happy life; more you cannot expect; a greater blessing the world cannot bestow. Busy you will certainly be, as the demand is great, both in private and public, for [those] with your training. Useful your lives must be, as you will care for those who cannot care for themselves, and who need about them, in the day of tribulation, gentle hands and tender hearts. And happy lives shall be yours, because busy and useful; having been initiated into the great secret—that happiness lies in the absorption in some vocation which satisfies the soul; that we have here to add what we can to, not to get what we can from, life.*

As a nurse and an analyst in information technology, my soul is satisfied from the work I've done and continue to do. My experiences at the bedside shape the way I approach any clinical tool we develop. I know I speak for those in my division and in the organization when I say that if our work saves one life, touches one family or makes the end-of-life process less painful for our patients and those they leave, then everything we've done has been well worth the effort. And for me, I am proud of the work we've done and those I've done it with. The IT world is ever-changing, and I don't know what will be around the corner, but I know as long as I work with those who have the patient needs at their core, the changes will only bring new and exciting solutions for clinical care providers and more comfort in some shape or form for our patients.

Reference

Osler, William (1999). 'Aequanimitas—The first essay.' The John Hopkins Health System—The John Hopkins University. Retrieved from www.medicalarchivves. jhmi.edu/osler/aequessay.htm (Accessed July 2018).

* * *

Our Voice: Healthcare Partner Council for Cleveland Clinic Information

Kathy Ray & Amy Szabo

Cleveland Clinic Information Technology Division has been able to engage patients directly into our daily work, staff meetings and projects.

I (Kathy Ray) am a Cleveland Clinic Regional Hospital Systems Analyst. Several years ago, I joined the Cleveland Clinic Main Campus Voice of Patient Advisory Committee (VPAC) to represent my mother. My mom is 83 years old, has multiple diseases and had been a patient at Main Campus, so I felt that I could represent her and be her voice. I participated in monthly meetings with other patients across the enterprise. We collaborated—sharing our patient stories with one another. A unique perspective to our VPAC was that not only did we have patients sitting at the table, but we also had leadership representation from all levels. The co-leaders were from the Office of Patient Experience. This was such a rewarding opportunity as I got to share not only the patient's side but also a caregiver's perspective as well.

As I participated on the Main Campus VPAC, my hospital president found out and asked me to co-chair a council at the regional hospital, and I did. Working with the Office of Patient Experience and Volunteer Services, we were able to invite eight patients to our council meetings. We always had leadership present for our meetings. We would provide food and a meeting room, and we would collaborate with one another. Often, we would have guest speakers attend and would begin gathering suggestions from our patients on how to improve service, processes and communications. It was at this time we received a new co-chair from the Office of Patient Experience, Amy Szabo. Amy was awesome and brought so much value to the group! The value added was a deeper connection between patients, and what mattered most to them, and caregivers/leaders.

VPAC was re-imagined with the voice of many patients and became Our Voice: Healthcare Partners program. Our Voice grew across the enterprise into many different councils. Amy and I were seeing such huge value in asking our patients to express themselves and offer solutions. It only seemed natural to spread Our Voice to the Information Technologies Institute. We presented the idea of forming an Information Technologies Healthcare Partner Council to Ed Marx and Maureen Sullivan, leaders of the Institute. Ed Marx championed the Our Voice mission, vision and values by engaging a patient to serve the Digital Global Advisory Council. That relationship was one that was sustained over time.

The IT Healthcare Partner Council kicked off with the strategy and forming session at a regional hospital with 35 caregivers and patients from across the enterprise. It was an interactive session where discovery and human-centered relationships were built. Collaboratively, what matters most to patients and caregivers was discussed in small groups and later shared with the larger group. As time passed, the relationships solidified and the demand for engagement increased. The Council is moving forward with engagement from across the enterprise, including distance relationships in other states.

The IT Healthcare Partner Council is embedding patients at the onset of design and project work in the following IT areas: Finance, electronic health record, digital platform, population health, imaging, clinical domain, security, infrastructure/network and project management.

When patients and families are empowered to become partners in all aspects of their healthcare, many improvements occur: Outcomes are better, patients and their families are happier and treatment can become more meaningful and understandable to patients.

Notes

1 https://psycnet.apa.org/doiLanding?doi=10.1037%2F0022-006X.61.4.678
2 https://onlinelibrary.wiley.com/doi/abs/10.1002/erv.1020
3 https://formative.jmir.org/2021/5/e27868/

Chapter 5

Communicate and Eliminate Barriers

Cross communication is essential to promote innovation.
By stripping virtual or physical barriers to communication,
ideas have a better chance of being realized.

Transparency is key to effective relationships, which are required for innovation to thrive. The depth and width you share will determine the size of your success. Look for every opportunity and platform to share while actively eliminating barriers to communication. Effective communication will make or break innovation. If active or passive resistance rises, so must your communication. While technology provides great tools to reach many, do not neglect the power of in-person eyeball-to-eyeball dialogue.

* * *

Enabling an Effective Remote Work Strategy

Sarah Hatchett

Introduction

Providing innovative, secure, and reliable technology is key to advancing our care delivery and experience for our patients and our caregivers. As leaders in healthcare technology, we have been called to enter the next phase

DOI: 10.4324/9781003372608-5

of our digital evolution during a period of major transition in healthcare and in spite of an unprecedented global pandemic. We have seen dramatic, dynamic changes in the way we work as teams—from the emergence of a remote workforce to the high demand for tools that enable and facilitate teamwork, IT has been at the center of enabling collaboration and engagement across our enterprise.

Since the start of the pandemic, we've been leading the way in innovating and adopting technologies within IT to help keep teams connected and productive. More than 90% of IT caregivers are now full-time work from home. We employ a "team of teams" approach for the nearly 950 caregivers in Ohio, Florida, and London. As we continue to grow globally as a "one ITD" team, we have had to shape our team culture intentionally to make sure everyone is engaged and included. Caregiver engagement rocketed from the 55th percentile to the 91st percentile in the midst of a highly stressful, fluid environment. It is truly amazing to consider this jump occurred at the same time we've significantly stretched our teams to help deliver initiatives that support both our COVID-19 response and our organizational growth mission. As a leadership team, we recognize that our teams thrive not only on collective focus on clear priorities but also in environments that enable the mission to be articulated consistently and the work to be delivered efficiently and effectively. The platforms and supporting tactics used to implement and sustain the remote work model have been absolutely essential to build our team identity and culture.

Now that remote work has been firmly established as our status quo, we have spent time reflecting on and sharing back the general best practices we've found to be successful.

Start at the Foundation by Defining What Your Goals Are for Your Remote Team

As a leader, you likely have something in mind for your ideal remote work culture. Don't leave it up to your teams to guess or even self-organize. Be specific in your direction, even if it is just to try and test something out for a short period of time. Make sure everyone understands the *why* of what you are trying to do, reiterate the message in multiple forums, and check for understanding.

For example, one of the first challenges we faced by going remote was the use of video conferencing technology. It wasn't until we cascaded intentional, directed messaging from our CIO that we began to see high levels of adoption. We equipped our leaders with a visual aid that show-cased the key benefits and encouraged everyone to insert it into the start of each meeting:

- ■ "Videos On" improves team engagement
  - o Increases focus on the meeting and participants
  - o "Puts a face to a name"—builds relationships
  - o Creates a connected culture
- ■ "Videos On" enhances interaction and learning
  - o Keeps meetings interesting and upbeat
  - o Visual information improves recall, retention, recognition
  - o Enables non-verbal communication and eye contact
- ■ "Videos On" improves productivity
  - o Fosters collaboration, builds trust
  - o Easier to make sure all questions and issues are addressed
  - o Positively impacts the meeting outcomes

Now we have team members encouraging each other to have videos on during meetings. We've had some caregivers report they are more familiar with their team members than when they were in the office setting due to video calls replacing dial-in only conference calls. By setting clear expectations and communicating consistently, we've been able to trans-form culture toward a model that promotes an environment we all want to be a part of.

Be Intentionally Human

In our remote work environment, it's very easy to seem two-dimensional and robotic, even distant. To your team, you literally are a talking head! In your approach as a leader, put effort into being empathetic, grateful, or even joyful at times. In the more stressful periods, even just being real with peo-ple about the situation helps us connect in a deeply earnest way. Your team needs to feel like you *see* them, and I don't just mean on camera. A great way to do this is to build in appreciation and recognition as a regular prac-tice. We have an easy-to-use enterprise-wide program called Caregiver Celebrations where you can call out caregivers who demonstrate outstand-ing behaviors and performance that align to our organizational values. There are leader boards for each business area to compete on who can send and receive the most recognitions, which serves as a motivating factor for some of our more competitive leaders! Even if you don't have a program like this, you could add a calendar reminder to help you send notes of appreciation habitually. Team members can also be acknowledged for work anniversaries, birthdays, or other special occasions. Sometimes it's as simple as briefly pausing at the start of a meeting to small talk or celebrate a holiday. All of these elements can help your teams connect through the screen and build meaningful relationships.

As with any major transformational change, remain sensitive and compassionate to team members that may be struggling. Take time to hear their concerns and needs, and maintain some level of flexibility to accommodate situations without having to compromise on your goals. Empower team members to solve their challenges creatively within the framework you lay out. As leaders, our goal is to create an environment that is both healthy and productive, which is only possible if each individual on the team is nurtured and guided toward that space.

Leverage Technology to Reduce Meeting Fatigue and Instead Engage Your Teams on Collective Priorities and Goals

Team members as humans need to feel that the work they do is meaningful and that they are directly contributing to the mission. This can be extremely hard to do when working remotely where the risk of burnout and feelings of isolation are at an all-time high. Whether it's through a formal methodology like OKRs or even just developing a visually clear roadmap of the work ahead, it's important to keep teams focused on the bigger picture and encourage them that what they do matters. Each month, our CIO addresses the entire global IT team via a virtual meeting to share key updates as well as vision and strategy. In addition, we try wherever possible to share videos of patient stories, have senior leaders showcase their personal and professional backgrounds, or invite guest speakers who highlight the impact of our work. We consistently see over 700+ attendees, which at more than 80% of the entire division demonstrates that the content is valuable and appreciated. Even in such a large meeting, we encourage engagement through real-time meeting feedback and chat features. These meetings and the supporting technology help build morale and motivation.

A major factor of burn out is the back-to-back nature of virtual meetings. Across our entire division we have implement a meeting management model to build consistency. This includes reducing standard durations from 60 minutes to 45 minutes in length, and avoid booking outside of business hours and during lunch. More recently, we've added a weekly Tuesday afternoon block period where recurring meeting series are discouraged to give team members more heads down time to get work done. Again, as we implemented, we have shared the key benefits of this model so everyone understands the purpose and goals:

- Shortened duration increases focus on meeting efficiency and productivity
- Allows gap time to record action items, respond to urgent messages, or take personal breaks
- Reduces fatigue and overscheduling

To make meetings interesting, we leverage Microsoft Teams platform capabilities like polls, whiteboards, GIFs, and more. Each leadership meeting we start with a virtual game—either an ice breaker or a simple challenge like trivia. With the explosion of remote workers across many industries, there are a ton of resources available online that can be leveraged. Consider additional training on the technology capabilities available to you—the more facile you can be with the platforms, the more creative you can get with facilitating engaging meetings that communicate your key messages.

Stay Classy and Have Fun

Everyone at Cleveland Clinic is a caregiver who provides direct patient care or supports those who do. However, as a remote workforce in a healthcare organization, we've had to promote mindfulness that not everyone we encounter virtually gets the benefit of working from home! As we work with our clinical caregivers, we have stressed the importance of maintaining professionalism and creating a culture of excellence around how work is executed and presented back to the business. This includes personal appearance but also the way meetings are conducted.

That being said, don't forget to have fun! Find ways to still engage in team building activities and create and maintain traditions. For example, in the early stages of the pandemic, while we were all in lockdown, I started using backgrounds on virtual meetings representing diverse locations as to take my teams on small 'trips.' I had them guess where in the world the photo was from before revealing the answer. I have covered cityscapes, natural wonders, works of art, stadiums, bridges, and more. This tradition continues to this day and adds a healthy amount of variety and levity to the daily routine. These fun elements make work enjoyable and help foster a sense of team identity.

Summary

As you work to develop your own best practices and remote work toolkit, establish an action plan, and consider taking measurements to evaluate what's working and what isn't. This could be through surveys, or it could be gathering feedback formally or informally. In IT we have formed our own Caregiver Experience council, which has been instrumental in evaluating and advising our senior leadership team on various tactics. This is a cross functional group of volunteers—both leaders and non-leaders—that have created a space to engage in conversation, share ideas, and provide a resource for implementing actions designed to continuously improve our workplace experience and satisfaction.

At the Cleveland Clinic, we are very lucky as leaders to have a clear, codified system of behaviors and values that are consistent across our enterprise. This model hasn't changed over time, and it still stays true even in our dynamically changing work environment. It's important to view your ability to thrive in a remote work setting as part of your leadership accountability. Implementing a successful remote work model requires leading change, driving results, inspiring and coaching, and connecting teams. In these times that can feel uncertain, fall back to these foundational capabilities as a guidepost to not only evaluate your own performance as a leader but give you support moving forward.

It can be difficult to overcome communication and collaboration barriers with a virtual, global model. We've recognized this remote work model is here to stay. Leaders need to be both aware and intentional about applying tactics and tools to improve engagement and enable collaboration across the enterprise. Through both the pandemic and our digital transformation journey, we have focused on expanding the best practices and tools we've developed in IT in order to become an innovative service partner, providing best-in-class technical solutions. Our enterprise collaboration tools and process have been a difference-maker in how we deliver the highest quality and compassionate care for our patients.

Together with our 70,000+ caregivers, our leadership in IT will continue to help Cleveland Clinic to be the best place to receive care—and to work—in all of healthcare.

* * *

The Technical Informaticist

John Lee

In medicine, there is a physiologic concept termed the "blood-brain barrier." As we have evolved, our bodies have created a physical and physiologic way to protect the very important brain and spinal cord. Much as the skull is a hard shell that protects your all-important brain from injury when you accidentally bump your head on a cabinet door as you stand up, this blood-brain barrier protects us from microscopic insults such as microorganisms and biochemical instabilities.

Although protecting the brain and spinal cord sounds like a good thing, sometimes this barrier can be problematic. For instance, if an infection does manage to set itself up in the central nervous system such as meningitis, delivering lifesaving antibiotics can be tricky. This makes the medication selection process more difficult. It is no longer as easy as choosing an antibiotic that eliminates a particular bacteria. One must consider the blood-brain physiology.

Our health IT systems are like our organizational central nervous system. To that end, it is common to create extra mechanisms to protect this central nervous system. For instance, it is common for healthcare IT departments to have security mechanisms that are far more stringent than nonhealthcare enterprises.

Similarly, many organizations are quite restrictive when giving access to the inner sanctums of IT configuration to only a privileged few with the "right" qualifications. Often, these qualifications are commonly determined based on a focus on IT requirements without as much consideration for domain expertise in operations or clinical content.

Unfortunately, this can create an unintended friction point as we try to deliver health IT projects to our end users. As our efforts have exploded since 2009's HITECH act, a trend has developed that many have noted can reduce that friction and facilitate solutions that can cross the IT blood-brain barrier.

I found myself inadvertently on the leading edge of that trend. In 2006, our group of emergency medicine physicians were tasked with converting our busy (50k+ visits/year) emergency department from paper to fully electronic. We unfortunately did not have the benefit of the support that comes with an enterprise "big bang" implementation. The emergency department was designated as the only department that was given this digital mandate.

I was designated as the physician subject matter expert, but that changed as I worked with the IT team. I became a bit impatient with tasks that I thought should be done faster. I also pushed back when I encountered resistance to concepts I suggested that I thought should be doable. As a result, I took it upon myself to learn as much about configuring the application as the analysts, even going so far to travel to the EMR headquarters to undergo technical training. Eventually, I became a hybrid physician-analyst. The result of this effort was an ED information system implementation that was quite successful.

A few years later, when our organization made the decision to move to Epic, I was determined to continue this effort. Unbeknownst to me, Epic had an entire community of like-minded physician informaticists that they termed "physician builders." Within the Epic community, there was an anecdotal understanding that organizations with such individuals performed better than their peers. For a while, this was a cultural understanding within the Epic community without robust evidence. Then, KLAS Research formed their Arch Collaborative in 2017. They surveyed various organizations on their success (or lack thereof) in creating EMR environments that are viewed positively (or negatively) by the end users.

They found a handful of strong positive correlations. One of the strongest was that organizations that had these "physician builders" were able to create content and environments that were highly statistically correlated with positive results (Davis & Bice, 2019).

If we focus on the underlying concept of a "physician" builder/analyst, the advantages extend to the larger clinical informatics team. In other words, one does not have to have an MD or DO in their title to fill this role. The key is that this informaticist needs to be someone who can translate between the clinical, operational and IT stakeholders. In fact, expanding the concept to nurses, pharmacists, therapists and other traditionally clinical front-line staff magnifies the effect. The right team members with the right skills can have a disproportionately positive effect. They effectively become force multipliers in the health IT effort.

If one considers this concept, the full power of a clinical informaticist is as a translator or Rosetta Stone. The expert clinical informaticist simultaneously knows how the organization operates, understands the clinical knowledgebase and has an expert grasp of the technical underpinnings of our digital systems. Such a person can competently attend operational meetings, speak to clinicians about the latest evidence-based practices and also be able to speak the technical lingo with an application analyst.

One does not have to be an expert in all domains. One informaticist can be a clinician primarily with expert knowledge on the latest medical evidence. Another can be an operational manager who is involved in day-to-day operations. Someone else can be an expert builder who can configure systems. The key though is that the informaticist has at least some expertise that crosses these normally siloed domains. They are able to cross our healthcare enterprise "blood-brain barriers" and be that special antibiotic that addresses the difficult to treat IT meningitis.

The expert clinical informaticist can often see opportunity where their clinical and operational colleagues don't know that a solution exists. In particular, my observation is that, in spite of a largely digital landscape, clinical operations still think in terms of paper. A common example is that of order sets. Most clinical end users still think of this content as monolithic paper-based content. However, the digital medium facilitates a dynamism that is not possible with paper. In our emergency department, we implemented order set content that dynamically morphed based on the clinical presentation of the patient. Because of this, our order set usage jumped dramatically and ultimately accounted for more than 60% of all orders entered in the emergency department. This content was not forced upon the users. They just gravitated to it because it was easier to use. From an operational and quality perspective, the benefit was that order variability was dramatically reduced. As an example, a nurse was more likely to see a limited number of pain medication protocols rather than the various versions that could be thought up by the ad-hoc imaginations of physicians (and we can be quite imaginative). Similarly, ordering clinicians would only see pediatric dosages for pain meds if the patient was a pediatric patient. Or, if the "pediatric" patient was a 200-pound high school football player, the adult dosages would automatically trigger. Certainly, this sort of content could have been generated by a more traditional IT-clinician dyad partnership, but it would have been much more difficult to identify the content and technical opportunities.

This is an example of the concept of "lowering activation energy." Those of you with a bent toward chemistry and biochemistry will recall that every chemical reaction needs some sort of nudge of extra energy to start the reaction. This is called the activation energy. There are external factors that can influence and reduce this activation energy. The chemical term for these factors are "catalysts." In biologic physiology, the proteins that act as catalysts are enzymes. In most instances, the biochemical, physiologic reaction cannot occur without these enzymes. In the world of health IT, the factors that lower this activation energy are clinical informaticists. The end result or project could potentially occur without them, but their presence makes the process so much easier and reduces waste and swirl.

I was a highly prolific physician builder, but the actual projects that I directly built are not the long-term value of the experience. Looking back, if I had remained a non-technical physician subject matter expert, I would not have nudged the IT teams to make many, if not most, of the key initiatives that benefited my organization. I attribute that to my experience in the builder role. Now I no longer build but am able to direct our constantly constrained resources better because I and similarly skilled informaticists can act as sociotechnical Rosetta Stones.

Ultimately, the role of such an informaticist is not to build large volumes of content. Rather, it is to convert the experience of building and configuring records into wisdom. That way the informaticist can better-guide and focus both the IT team and our operational partners to ideate and govern what can and should be developed. They are IT enzymes.

This team of informaticists was able to deliver key accomplishments. For example, during Joint Commission inspections, our PRN (as needed) pain medication instruction compliance was noted to be some of the best they had seen. The team was able to revamp order set governance and structural design changes, resulting in reduced order set build time by more than 50%, systematized standardization and improved quality. Our clinical alerts went from a response rate of 6% to 9%, while total alerts per provider were reduced by 28%. The skill level of the clinical informatics team reduced the "activation energy" and made these results far more accessible than without their efforts.

One can achieve such results without physicians and other clinical staff learning the technical underpinnings of your system, but having such skills does make it easier. As I mentioned, they become force multipliers. If done right, their value far exceeds their FTEs or salary number. You do need to find the right people and point them in the right direction. On top of that, governance is really important and in fact is also part of the long-term benefit. You will groom leaders who understand the reasons for such concepts such as governance and the difficult efforts related to governance. Having such discussions with a clinical colleague is very different than getting it from an IT staffer whom you don't think "gets it." Likewise, if you are an IT person, having someone who has technical chops and can explain the technical path in terms that you understand is really helpful. That is the utility in having these team members who can lower the activation energy and cross that blood-brain barrier.

Reference

Davis, T., and Bice, C. (2019). *Arch Collaborative Guidebook*. https://klasresearch.com/archcollaborative/report/arch-collaborative-guidebook-2019/293. Accessed 9/23/22.

* * *

How Can Health Systems Leverage Technology to Engage Patients, Especially Those Who May Lack Digital Skills, Smartphones, and Robust Internet and Mobile Connectivity?

Amy R. Sheon & Leslie Carroll

At a recent health innovation event, I (ARS) asked a distinguished panel of health system executives the following questions:

> *Patient engagement has been deemed the blockbuster drug of the millennium. How should innovators address patient engagement? And what about the patients, such as the 48% of low-income Cleveland residents who lack any internet subscription or the 44% of all Cleveland residents who lack a mobile data plan? [1] What about residents who have broadband and mobile data plans but lack the skills needed to use them?"*

The answers, such as promoting empathy among providers, were unsatisfyingly health system–centric. With careful thought, health systems may see consumer health technology adoption move faster than the 17-year gap that characterizes adoption of other health innovations [2, 3].

How Will That Unfold?

Health tech firms will widely publicize compelling anecdotes such as a health system that avoided a $12,000 readmission penalty when a patient dashboard alerted Mary's provider that her daily weight measurement—transmitted automatically from her Bluetooth-enabled scale—revealed a sudden weight gain, leading to rapid adjustment of medication for her congestive heart failure. Health systems and insurers will widely promote use—and even subsidize patient purchase of such technology—but be disappointed that the needle does not move on the desired outcome in the subsequent two years. Delving into their analytics, health systems will note very low uptake of patient-focused technology, especially among populations that experience challenges in using digital health technology. A couple of more years of research will reveal other causes of lack of widespread adoption by patients, such as biases in who received encouragement from providers to adopt technology and recognition that patients need help in using the technology and in understanding its value and that some populations face unique challenges with technology use. Examples include patients with low vision being unable to read small screens, or those with limited dexterity having difficulty entering information on small keypads. Health systems will then have nurses, techs, and even physicians instruct patients in how to download apps and operate digital devices. A couple of more years down the road, health systems will find that training will have helped patients who are already adept at using technology with adopting digital health tools but will have left behind the substantial fraction of patients deemed "digitally unready" by Pew Research [4]. Such individuals will still have resisted adopting telemedicine, remote monitoring, or using apps to track their health or report health measures.

Within a decade from now, solid evidence will point to the disastrous impact of the digital divide on the ability of patients to manage not only their health but also to address the factors that make the largest contribution to their health, their social determinants of health—such as education, employment, and housing. Healthcare only accounts for approximately 20% of differences in health outcomes, according to models widely accepted by the Institute of Medicine and documented in public health studies [5–10]. The enormous variability in disease and risk factors by zip code, race, and income [11–14] underscores the overall limited impact of healthcare on population health and on the disparities that hold back overall improvement.

So What Is a Health Care System to Do?

Patient portals to electronic health records offer obvious benefits in the way of convenience (appointment scheduling, checking test results, requesting prescription renewals, secure messaging). They also serve as a conduit for high-quality health information and can offer the ability to track health conditions and report health outcomes. Portals are ubiquitous and applicable to all health conditions, yet they can be personalized to address language and literacy barriers. Instruction in use of a portal could reap a lifetime of dividends insofar as portals could be considered a gateway to other connected health tools, including telemedicine, remote monitoring, apps, wearables, and more.

Efforts of healthcare systems to address risk factors associated with disease, such as diet, exercise, smoking, and tobacco, are growing with healthcare's use of Screening, Brief Intervention and Referral to Treatment (SBIRT) [15]. Innovation can surely help with the difficult task of matching patients to community resources that are convenient and culturally relevant, an approach that we are beginning to call Precision Public Health [16]. To potentiate this approach, however, patients must have ready, secure, and continuous access to private electronic communications. Health systems can be powerful advocates by screening patients for their digital skills and connectivity, referring patients to community organizations that can address gaps, and "prescribing" patient utilization of portals and other such technology.

Local Experience: MetroHealth System, Cleveland, Ohio

We believe there is much to be learned from our own local efforts to promote patient adoption of portals. As Cleveland's only public health care provider, the MetroHealth System had nearly 1.2 million outpatient and 125,000 emergency department visits in 2016. Despite being an early and nationally recognized [17] leader in the adoption of electronic health records, only 29.1% of patients at our local public hospital, MetroHealth System, had initiated patient portal use by 2015, with significant disparities seen by race (23.4% for blacks, 23.8% for Hispanics versus 34.1% for whites) and insurance (17.4% adoption for uninsured patients versus 39.3% for those with commercial insurance) [18]. In multivariate analysis, the strongest predictor of portal use, controlling for patient demographic characteristics, was neighborhood-level access to broadband internet [18]. It may be tempting to assume that lack of internet subscription reflects lack of interest in subscribing. Yet lack of interest accounts for just more than one-third (34%) among the 11% of adults nationwide who do not use the internet. Nearly the same fraction (32%) cite difficulty as the main reason for non-use, and 19 percent cited cost as the main barrier [19].

In conjunction with efforts to obtain federal approval for mergers among large telecommunication companies and cable providers, ISPs in several states began offering, in 2016, discount unlimited high-speed broadband internet to individuals meeting various eligibility criteria, such as receiving SNAP food assistance or Medicaid. (Requirements vary by service provider and by state.) or low-income residents [20, 21] Our local community partners, the Ashbury Senior Computer Community Center (ASC3), and Connect Your Community began referring digital skill training clients to these ISPs, but many clients were told they were ineligible for the discount broadband rate *because high-speed broadband was not available in their neighborhoods.* Looking into this seeming catch-22, our partners were startled to discover what appeared to be digital redlining—an ISP that "systematically discriminated against lower-income Cleveland neighborhoods in its deployment of home Internet and video technologies over the past decade" [22]. Upgrades to broadband fiber had been made in suburban and middle-income urban Cleveland neighborhoods, but the "overwhelming majority" of city blocks with high rates of poverty were left with internet access speeds too slow to allow for signing into portals without timing out. (A similar pattern of digital redlining was found by others for the entire State of California [23] and by our colleagues in other Midwest cities [24, 25].)

Thus, it is important for providers not to withhold portal recommendations based on an assumption that some populations are less likely to use the internet than others, or that non-use reflects lack of patient interest.

The Cleveland Digital Patient Engagement Model

Confronted with low rates of portal adoption and limited access to internet, we developed a multipronged model system that contains elements that we believe could benefit nearly all health systems. The model consists of the following components:

1. Health systems identify patients not currently using portals, preparing a daily list for the clinic.
2. CHWs screen non-portal users for digital skills and connectivity during a very brief discussion at the end of a patient appointment.

3. CHWs refer patients lacking skills or connectivity to local community organizations for assistance. This referral is more successful when the CHW follows up or even provides some training on the spot.
4. Once they have attained basic digital skills and connectivity, CHWs then directly train patients to use the portals either at the end of the visit or at a later scheduled time.

Development of our program required a number of precursor steps, as summarized in Figure 5.1a. We developed our program in partnership with ASC3, and the local United Way 2-1-1 referral service to identify additional training opportunities such as at libraries, job training centers, and public housing communities.

1) Modify workflow if needed to ensure provider encouragement of patient portal use.
2) Develop list of digital skill training and connectivity support resources.
3) Create referral partnerships with local digital inclusion advocates such as libraries and employment training facilities for basic digital skill training and connectivity support.
4) Recruit and train CHWs to screen and refer patients to these resources.
5) Train CHWs to use literacy acquisition strategies to teach patients to use portals.
6) Develop reports to track portal use patterns and disparities.

1) Adopt universal screening of patients for digital skills and connectivity.
2) Adopt opt-out policies for portal creation.
3) Work with portal vendors to improve the portal user experience, especially regarding authentication.
4) Generate evidence for digital medicine.

Figure 5.1 (a) **Preparatory steps for digital patient-engagement model implementation.** (b) **Additional recommended steps to optimize patient portal use and accelerate patient HIT adoption.**

As is the case in many communities, libraries and "digital inclusion" organizations are experts in assessing digital skills and addressing skill gaps. The National Digital Inclusion Alliance maintains an online listing and map of local affiliates around the country (digitalinclusion.org). The United Way 2-1-1 referral line also geared up to refer patients by phone to the nearest location to obtain digital skill training for free or low cost, have access to refurbished computers and mobile phones, and enroll in discount internet access programs.

Community Health Workers as Key Drivers of Digital Patient Engagement

Community Health Workers (CHWs) are an optimal workforce to help healthcare organizations understand and address patient barriers to technology adoption and to help patients attain the digital skills and connectivity needed to use health IT [26]. A 1998 study identified optimal CHW qualities and roles [27] that have been widely incorporated into state and national standards and certifications [28–32]. Recommended CHW qualities that support this effort include: Being a member of or having intimate familiarity with the lived experience of vulnerable patients and being friendly, nonjudgmental, empathetic, a lifelong learner, persistent, resourceful, and committed to community improvement [27].

Although not adopted in our program, other practices outlined in Figure 5.1b were suggested [33–35] by participants at a recent national gathering of digital literacy experts meeting together with those promoting health portal adoption [36].

Figure 5.2 shows a crosswalk between CHW roles and tasks needed to promote patient digital engagement.

■ *Modify workflow if needed to ensure provider encouragement of patient portal use. Providers may be reluctant to encourage patients to use portals, fearing being overloaded by patient messages. Office personnel should triage these messages, which should reduce overall provider burden. Adopt universal screening of patients for digital skills and connectivity and referral to local digital inclusion advocates to address gaps.* Recognizing that healthcare isn't necessarily the most urgent priority in the lives of many patients, healthcare institutions are increasingly adopting routine screening and referral of patients to address social

Cultural mediation among individuals, communities, and health systems	• Helps health care providers understand patient barriers to HIT use • Helps patients understand the value of HIT, address barriers such as trust and privacy
Providing culturally appropriate health education and information	• Helps patients understand information in the portal or find additional information about their condition
Providing coaching and social support	• Encourages patient self-confidence with using technology • Engage family or other caregivers to support patient digital engagement or to access HIT as proxies
Advocating for individuals and communities	• Raise awareness about social justice dimension of disparities in digital skills and connectivity • Encourage engagement of patients in technology design to increase usability
Building individual and community capacity	• Empower patients through increased digital and health literacy and connectivity • Strengthen relationships between health care institutions and local digital inclusion organizations
Providing direct service	• Screen patients for digital literacy and connectivity • Refer patients to local partners to address gaps • Teach patients to use digital health tools
Participating in research and evaluation	• Contribute to study design, data collection, report writing and presentations about disparities in digital patient engagement, efforts to address gaps, the impact of digital engagement on health care quality and outcomes, etc.

Figure 5.2   CHW roles and digital engagement duties.

Source: Rosenthal et al., 2016.

determinant of health barriers, such as lack of housing, domestic violence, or emergency income needs. For example, in the Accountable Health Communities Model, the Centers for Medicare and Medicaid Services is testing such a program in 30 cities across the country, including Cleveland, where the local United Way 2-1-1 referral program is working with MetroHealth System. This is a missed opportunity to assess digital skills. Patients lacking skills, equipment, broadband, and/ or mobile data plans should be referred to local organizations focused on digital literacy and access. Public libraries and job training organizations address the skill training, but connectivity is an important gap, especially in communities lacking competition to drive down the price of mobile and broadband internet service [37].

▪ *Adopt opt-out policies for portal adoption.* Countless studies have shown a large drop off in the number of patients who initiated versus continue to use portals. Many drop off after the initial attempt, which may not even lead to a successful login. One system found that going to opt-out policies increased overall adoption and reduced disparities in portal use [28].

▪ *Address technology usability.* The poor usability of consumer-facing health IT has been widely lamented. Gibbons et al. found that including individuals who experience technology challenges at the earliest stages of technology development [38] leads to technology that is better adapted to all populations. A local program that trains patients to collaborate in research [39] could train patients to provide technology usability feedback to developers.

■ *Generate evidence for digital medicine.* Unlike drugs and devices that require FDA approval, there are no standards or regulations to ensure that digital health technologies are safe and effective. NODE Health, the Network of Digital Evidence for Health, is promulgating such standards for digital medicine and creating a registry of technologies to help speed development and adoption of effective technology [40]. In collaboration with other NODE Health leaders, and with funding from Case Western Reserve University's Clinical and Translational Science Collaborative (NIH award UL1TR000439–10), has published recommendations for ensuring that digital medicine technologies address populations that have special needs with respect to use of digital health technology [41]. Health systems should participate in studies of digital health technology, paying special attention to the needs of patients who may lack digital skills and robust connectivity.

Conclusion

The $28 billion federal investment in health information technology, plus general advances in technology and cloud computing, have created a robust and rapidly growing ecosystem of technology-based tools. Disparities in digital skills and connectivity portent that such technology will accentuate vast and growing health disparities. Health systems can prevent this otherwise predictable result through conscious effort across the full spectrum of the health technology ecosystem. Engage in research on technology development and ensure that diverse populations contribute to all stages of product development. Systematically screen all patients and refer them to local digital inclusion advocates to attain the general digital skills and connectivity needed before they can be encouraged to adopt digital technology. Health systems should engage community health workers to perform this screening and referral, and then train patients, once suitably equipped, to use digital health technology. By adopting such measures, health system executives will have compelling answers to questions about how to support patient engagement.

References

1. US Bureau of the Census. Types of computers and internet subscriptions 2017 American community survey 1-year estimates, Table S2801. [cited 2018 Dec 6]. Available from: https://factfinder.census.gov/bkmk/table/1.0/en/ACS/17_1YR/S2801/1600000US3916000

2. Balas EA, Boren SA. Managing clinical knowledge for health care improvement. *Yearb Med Inform.* 2000;(1):65–70.

3. Morris ZS, Wooding S, Grant J. The answer is 17 years, what is the question: Understanding time lags in translational research. *J R Soc Med.* 2011 Dec [cited 2018 Apr 18];104(12):510–20. Available from: www.ncbi.nlm.nih.gov/pmc/articles/PMC3241518/

4. Horrigan JB. Digital readiness gaps. *Pew Research Center.* 2016 Sep [cited 2017 Feb 14]. Available from: www.pewinternet.org/2016/09/20/digital-readiness-gaps/

5. Braveman PA, Egerter SA, Mockenhaupt RE. Broadening the focus: The need to address the social determinants of health. *Am J Prev Med.* 2011 Jan;40(1, Supplement 1):S4–18. Available from: www.sciencedirect.com/science/article/pii/S0749379710005635

6. Commission to Build a Healthier America. *Breaking Through on the Social Determinants of Health and Health Disparities: An Approach to Message Translation.* Robert Wood Johnson Foundation, 2009. Available from: www.rwjf.org/en/library/research/2011/05/breaking-through-on-the-social-determinants-of-health-and-health.html

7. Marmot M, Friel S, Bell R, Houweling TA, Taylor S, Commission on Social Determinants of Health. Closing the gap in a generation: Health equity through action on the social determinants of health. *Lancet.* 2008;372(9650):1661–9. Available from: http://whqlibdoc.who.int/publications/2008/9789241563703_eng.pdf?ua=1

8. Whitehead M, Dahlgren G. What can be done about inequalities in health? *Lancet.* 1991 Oct [cited 2018 Apr 12];338(8774):1059–63. Available from: http://linkinghub.elsevier.com/retrieve/pii/014067369191911D

9. Williams DR, Costa MV, Odunlami AO, Mohammed SA. Moving upstream: How interventions that address the social determinants of health can improve health and reduce disparities. *J Public Health Manag Pract.* 2008 Nov [cited 2018 Feb 23];14(Suppl):S8–17. Available from: www.ncbi.nlm.nih.gov/pmc/articles/PMC3431152/

10. Woolf SH, Braveman P. Where health disparities begin: The role of social and economic determinants—and why current policies may make matters worse. *Health Aff (Millwood).* 2011 Oct 1 [cited 2011 Oct 12];30(10):1852–9. Available from: http://content.healthaffairs.org/content/30/10/1852.abstract

11. Mokdad AH, Ballestros K, Echko M, Glenn S, Olsen HE, Mullany E, et al. The state of US Health, 1990–2016: Burden of diseases, injuries, and risk factors among US states. *JAMA.* 2018 Apr 10 [cited 2018 Apr 11];319(14):1444–72. Available from: https://jamanetwork.com/journals/jama/fullarticle/2678018

12. Case Western Reserve University. *Cleveland & Cuyahoga health data matters.* 2018 [cited 2018 Jan 23]. Available from: https://hdm.livestories.com/

13. Centers for Disease Control and Prevention. *500 cities project: Local data for better health*. [cited 2017 Jul 14]. Available from: www.cdc.gov/500cities/

14. Robert Wood Johnson Foundation and the University of Wisconsin Population Health Institute. *County Health Rankings & Roadmaps*. 2017 [cited 2013 Nov 12]. Available from: www.countyhealthrankings.org/app/ohio/2017/rankings/cuyahoga/county/outcomes/overall/snapshot

15. Agerwala SM, McCance-Katz EF. Integrating screening, brief intervention, and referral to treatment (SBIRT) into clinical practice settings: A brief review. *J Psychoactive Drugs*. 2012 [cited 2018 Apr 13];44(4):307–17. Available from: www.ncbi.nlm.nih.gov/pmc/articles/PMC3801194/

16. Frank S. Precision public health is why health data matters. *Cleveland and Cuyahoga Health Data Matters*. 2016 [cited 2018 Apr 13]. Available from: www.healthdatamatters.org/blog/2016/11/22/precision-public-health-is-why-health-data-matters-blog-by-scott-frank-md-ms-november-2016

17. Kaelber D. Exploiting health information technology to improve health: The MetroHealth 2015 Davies award enterprise application. *Healthcare Information and Management Systems Society (HIMSS)*. 2015 Aug [cited 2017 Feb 4]. Available from: www.himss.org/metrohealth-cleveland-davies-enterprise-award?ItemNumber=46432

18. Perzynski AT, Roach MJ, Shick S, Callahan B, Gunzler D, Cebul R, et al. Patient portals and broadband internet inequality. *J Am Med Inform Assoc*. 2017 Mar 23 [cited 2017 Apr 14]. Available from: https://academic.oup.com/jamia/article/doi/10.1093/jamia/ocx020/3079333/Patient-portals-and-broadband-internet-inequality

19. Anderson M, Perrin A, Jiang J. 11% of Americans don't use the internet. Who are they? *Pew Research Center*. 2018 [cited 2018 Apr 13]. Available from: www.pewresearch.org/fact-tank/2018/03/05/some-americans-dont-use-the-internet-who-are-they/

20. AT&T Low-Cost Internet Program. 2016 [cited 2016 Apr 2]. Available from: http://connectyourcommunity.org/wp-content/uploads/2016/03/LCBB_SNAP_OnePager_020116.pdf

21. Callahan B. CYC 2.0 calls for digital inclusion investment as price of Comcast-Charter deal. *Connect Your Community 2.0*. 2014 [cited 2016 Mar 12]. Available from: http://connectyourcommunity.org/cyc-2-0-calls-for-digital-inclusion-investment-as-price-of-comcast-charter-deal/

22. Callahan B. AT&T's digital redlining of Cleveland. *National Digital Inclusion Alliance*. 2017 [cited 2017 Apr 16]. Available from: https://digitalinclusion.org/blog/2017/03/10/atts-digital-redlining-of-cleveland/

23. Strain G, Moore E, Gambhir S. AT&Ts digital divide in California. *Haas Institute, University of California, Berkeley*. 2017 Apr [cited 2017 Apr 26]. Available from: http://haasinstitute.berkeley.edu/digitaldividecalifornia

24. Callahan B. Map of Montgomery County, OH census blocks with AT&T "Fiber To The Node" (VDSL) Internet access at 18 mbps or more. *National Digital Inclusion Alliance*. 2017 [cited 2017 Apr 16]. Available from: https://digitalinclusion.org/montgomery-county-att-fttn-available/

25. Callahan B. More digital redlining? AT&T home broadband deployment and poverty in Detroit and Toledo. *National Digital Inclusion Alliance.* 2017 [cited 2018 Apr 13]. Available from: www.digitalinclusion.org/blog/2017/09/06/more-digital-redlining-att-deployment-and-poverty-in-detroit-and-toledo/

26. Sheon A. The ROI of digital inclusion for community health and health care [Internet]. *National Digital Inclusion Alliance Net Summit.* 2017 May 17 [cited 2017 Apr 18]. Available from: https://digitalinclusion.org/wp-content/uploads/2017/04/NDIA-2017-FINAL.pptx

27. Rosenthal EL, Rush CH, Allen CG. *Understanding Scope and Competencies: A Contemporary Look at the United States Community Health Worker Field.* University of Texas—Houston School of Public Health, Institute for Health Policy. 2016 Apr [cited 2018 Apr 20]. Available from: www.healthreform.ct.gov/ohri/lib/ohri/work_groups/chw/chw_c3_report.pdf

28. *Community Health Workers: Action Guide for CHW Employers.* ICER. [cited 2018 Jan 30]. Available from: https://icer-review.org/material/chw-action-guide-employers/

29. *Community Health Workers: Action Guide for Workforce.* ICER. [cited 2018 Jan 30]. Available from: https://icer-review.org/material/chw-action-guide-workforce/

30. Mejia-Rodriguez C, Spink D. Training curriculum for community health workers [Internet]. [cited 2018 June 30]. Washington State Department of Health. Report No: DOH 140–043. 2015 Aug. Available from: www.doh.wa.gov/Portals/1/Documents/Pubs/140-043-CHWT_ParticipantManual.pdf

31. Findley SE, Matos S, Hicks AL, Campbell A, Moore A, Diaz D. Building a consensus on community health workers' scope of practice: Lessons from New York. *Am J Public Health.* 2012 Oct [cited 2017 Aug 4];102(10):1981–7. Available from: www.ncbi.nlm.nih.gov/pmc/articles/PMC3490670/

32. Berthold T, ed. *Foundations for Community Health Workers.* 2nd edition. Jossey-Bass; 2016: 736.

33. Ancker JS, Barrón Y, Rockoff ML, Hauser D, Pichardo M, Szerencsy A, et al. Use of an electronic patient portal among disadvantaged populations. *J Gen Intern Med.* 2011;26(10):1117. Available from: https://link.springer.com/article/10.1007/s11606-011-1749-y

34. Peacock S, Reddy A, Leveille SG, Walker J, Payne TH, Oster NV, et al. Patient portals and personal health information online: Perception, access, and use by US adults. *J Am Med Inform Assoc* [Internet]. 2017 Apr 1 [cited 2017 Mar 21];24(e1):e173–7. Available from: https://academic.oup.com/jamia/article/24/e1/e173/2631484/Patient-portals-and-personal-health-information

35. Ancker JS, Nosal S, Hauser D, Way C, Calman N. Access policy and the digital divide in patient access to medical records. *Health Policy Technol.* 2017 Mar [cited 2017 Apr 25];6(1):3–11. Available from: http://linkinghub.elsevier.com/retrieve/pii/S2211883716300867

36. Harris K, Sheon A, Castek J, Perzynski A, Reeder J, Sieck C. Workshop: Multidisciplinary perspectives on digital adoption. 2018 Apr [cited 2018 Apr

14]. Available from: https://netinclusion2018.sched.com/event/DKpF/multidisciplinary-perspectives-on-digital-inclusion-and-health

37. *Strategies and Recommendations for Promoting Digital Inclusion.* Consumer and Governmental Affairs Bureau, Federal Communications Commission. 2017 Jan. [cited 2018 June 30]. Available from: http://transition.fcc.gov/Daily_Releases/Daily_Business/2017/db0126/DOC-342993A1.pdf

38. Gibbons MC, Lowry SZ, Patterson ES. Applying human factors principles to mitigate usability issues related to embedded assumptions in health information technology design. *JMIR Hum Factors.* 2014 [cited 2017 Apr 29];1(1):e3. Available from: http://humanfactors.jmir.org/2014/1/e3/

39. Theurer J, Pike E, Sehgal AR, Fischer RL, Collins C. The community research scholars initiative: A mid-project assessment. *Clin Transl Sci.* 2015;8(4):341–6.

40. Atreja A, Bates D, Clancy S, Daniel G, Doerr M, Franklin P, et al. *Mobilizing Mhealth Innovation for Real-World Evidence Generation.* Duke University Margolis Center for Health Policy; 2018.

41. Sheon AR, Van Winkle B, Solad Y, Atreja A. An algorithm for digital medicine testing: A node. Health perspective intended to help emerging technology companies and healthcare systems navigate the trial and testing period prior to full-scale adoption. *Digital Biomarkers.* 2018 [cited 2018 Dec 7];2:139–54. Available from: www.karger.com/Article/FullText/494365

* * *

When Innovation Meets Integration: Developing the Tools for Invisible Healthcare

Drew Schiller

Two hundred forty million Americans own a smartphone [1], and we spend an average of five hours each day interacting with our devices [2]—nearly one-third of our waking hours. The expansion and mass adoption of these mobile computers over the last decade has implications for how healthcare will function to interact with patients and provide better care in coming years. Perhaps more importantly, the evolution of smartphone technology itself offers a roadmap for how healthcare can integrate novel technology to design an *invisible*, interoperable healthcare system.

The importance of system interoperability is an ever-present, if not prosaic, mantra for many of us in healthcare. Embedded systems must seamlessly exchange information with other technologies inside and outside the clinical walls for both clinicians and patients. Yet, the concept of invisible design, which can be considered *applied interoperability*, is a fairly new construct for healthcare.

Invisible design means something is so well integrated into your routine that you don't notice the mechanics of it, only the outcome. Invisible tools become a natural part of your day-to-day life without consciously thinking about it.

Smartphones, for example, have capitalized on this concept of designing for an invisible experience. With voice-activated assistants, you no longer have to look at your device to access functionality. You can ask Siri, Alexa, or Google to play music, schedule a meeting, or even send a text without opening a single application. You can program your weather or news app to send you a synopsis each morning to prepare you for the day. You can queue your music or maps to play as soon as you enter your car. We no longer have to search or monitor for relevant information. By design, our smartphone pushes it to us.

The concept of invisible design was best summarized by user interface guru Jared Spool, who said, "Good design, when it's done well, becomes invisible. It's only when it's done poorly that we notice it."

And in healthcare, we certainly do notice it. When we talk to patients, providers, and healthcare administrators about the design of the healthcare experience today, we find that challenges and faulty mechanics are painfully obvious to all stakeholders, especially to the patient, a role we all share.

One such patient, Steve, was diagnosed with type 2 diabetes at age 40, like his father, uncle, and grandmothers before him. A decade after his diagnosis, Steve continued to struggle. He struggled to maintain a healthy weight despite working out and leading an active lifestyle. He struggled with his a1c levels despite a managed diet and nutrition program. He struggled with depression and energy levels despite cutting his work hours to part-time. Even with weekly office visits, medication, and an established treatment program, Steve struggled to feel in control of his condition and overall health—and clearly not for a lack of trying.

Steve was tirelessly dedicated to getting better, so much so that he would routinely fax a copy of his glucose readings—which he kept in an excel file—to his doctor ahead of each visit. But, with no clinical protocol or system in place to manage Steve's data, the endocrinologist would promptly discard Steve's readings and instead continue to manage his condition using only preceding a1c readings. This caused a frustrated Steve to leave his provider and seek treatment with a new kind of program, one that would integrate the data he generated daily to better manage his condition.

Steve found and enrolled in a new program of care, which, unlike any program Steve had been a part of before, leveraged his biometric and observations of daily living (ODLs) data to manage his condition. The program required Steve to take blood glucose readings at least three times a day, usually around meals. Steve's data were collected via his program-issued application, which connected data from his weight scale, a blood glucose meter, an activity tracker, and a blood-pressure monitor (to help monitor the effect of medication on his kidney function).

Steve's continuous data sharing, combined with a programmatic monitoring of his data, revealed an overlooked snacking habit that affected his ability to control his glucose levels. While watching television with his kids each night, Steve would eat a couple of handfuls of chips or popcorn (which Steve thought were "free" because of the relatively few calories he consumed). The late-night snacking sessions caused a consistent spike in Steve's blood sugar each morning. Steve's care manager noticed the trend and informed him on how to amend his diet and actions (corn is high in carbohydrates, and Steve learned that this turns into sugar in his body). As this demonstrates, by employing patient-generated health data (PGHD) in a consistent feedback loop, a previously unidentified and problematic trend was identified, and the patient was able to adjust his lifestyle to achieve his target HbA1c.

Today, Steve is down 50 pounds and has lowered his a1c from 8.9 to 6.5. Additionally, after showing success in consistently managing his weight and condition, Steve's physician amended the number of in-office visits he needed to once every three months. Reviewing his data with a clinician, being accountable for measurements, and building a rapport with a care team helped Steve gain deeper insight into his condition and learn how to better self-manage.

For patients like Steve and the millions more suffering from chronic conditions, the inability to share data with a physician was thwarting better outcomes. His traditional program of care focused solely on his a1c levels, viewed retrospectively every three months. Despite monitoring several aspects of his health multiple times a day, he was not able to leverage this information with his care team to impact his treatment.

So, what did Steve do differently?

In reality, very little. Steve changed almost nothing about his routine. Simply integrating the data he was already generating into the clinical workflow with a defined program of care changed this man's life. The program enabled superior patient–provider discourse by changing the conversation from, "Why isn't this working?" to "Here's what the data show is not working." To do this, the program did not implement new or ground-breaking technology. The program simply integrated existing information, devices, and clinical systems to yield the improved outcomes.

Oftentimes, true innovation comes from restructuring current technologies in new and interesting ways. Take, for example, the iPhone, released in 2007, and its predecessor, the PalmPilot, released in 1997. The PalmPilot was a bulky smart device with a stylus that offered basically the same functionality as the first-generation iPhone. In fact, there was little innovation in the features offered by Apple's new smart device. Much like the iPhone, the PalmPilot offered phone services, text messaging, a calculator, email, games, a QWERTY keyboard, and a web browser. Apple didn't focus on developing new functionalities for a device; they focused on taking these applications and embedding them within a more seamless, more *invisible* user experience. The focus on invisible design clearly paid off—the phone was an instant hit, and since the launch of the iPhone, Apple has become among the most regarded and valuable companies in the world.

To further enhance the user experience, Apple not only changed the way phones were designed, they also changed the way phones were integrated with carriers. Prior to the iPhone, carriers were white-labeling cell phones and adding their own software. Apple rejected this standard, forming an exclusive launch with AT&T to release the iPhone with Apple's iOS installed and untouched. They wanted to maintain control over all aspects of the iPhone's user experience—because, candidly, consumer electronics companies deeply understand what healthcare has not yet internalized: How important the experience is to engagement.

In 2008, a year after launching the iPhone, Apple released the App Store. It was no longer solely Apple dictating the user experience. Consumers could now personalize their device to the experience *they* wanted.

Now, more than a decade later, smartphones are seemingly sewn to the hands of consumers. With each year, these devices are better designed to integrate directly into our personal experiences—helping us capture, share, and manage our lives.

Similar to 2007 with the launch of the iPhone, healthcare today has the technology—the feature set—to design a better system. However, we have yet to integrate these technologies in such a way that facilitates seamless, personalized user experiences. To move past the current stagnation of the industry, we must apply innovation to the integration of technology to create unobtrusive, invisible healthcare experiences.

In 2017, the Commonwealth Fund conducted an evaluation of the U.S. healthcare system along with 11 similar countries. The group found that the

U.S. system performed the worst among the countries, while spending more. Particularly, the U.S. system performed exceedingly poor on most, if not all, measures related to population health, ranking last in affordability, access, and outcomes [3].

This evaluation elaborates on something many of us know to be true: American patients today are struggling to find available, affordable healthcare that truly helps them achieve their health goals.

> "It costs too much."
> "It took forever for the doctor to see me."
> "I can't access my health record to send to my new physician."
> "I have gone to three different doctors, and they still don't know what's wrong."

These are common refrains heard from patients like Steve, echoing frustrations with their healthcare experiences. Even as regulatory measures offer guidance, penalties, and incentives to drive the industry toward value-based care, barriers to achieving improved access, affordability, and outcomes remain.

To do this, we need to address two critical shortcomings of the current system's design. First, we need data-driven technology to be integrated into a comprehensive program of care that personalizes the treatment and experience of each patient. Second, we need for this technology to be integrated into the providers' workflows to make the experience of delivering personalized care easy for clinicians.

Data-driven programs, like the one designed for Steve, are the key to a results-driven, invisible healthcare experience. Having a technology-enabled program of care is crucial in allowing care teams, like Steve's, to identify negative progressions and elevate the right data for intervention and engagement. When clinicians are able to see patient-generated data—in real time, in their existing workflow—they are able to more efficiently identify trends that could indicate a looming negative health event. In Steve's case, the care team was able to alert the physician of concerning data, initiating a life-changing conversation. Steve had been tracking these events throughout the management of his condition, but it was not until the data was made available to a care team, actively monitored, and proactively used in care that changes were able to be made to Steve's lifestyle.

Steve does not have the education of the healthcare professional. He needed a diabetes educator to guide his nutrition and activity, a nurse care manager to monitor his treatment, and an endocrinologist to adjust medication based on a trend of glucose and activity values rather than one a1c reading every three months. And, in return, the care team needed access to Steve's data to provide the right education, monitoring, and treatment to affect his condition. They needed this access provided in the workflow with alerts and triggers to notify when intervention is needed. Both the patient and the provider needed pathways for better communication. For Steve and his care team, data provided an objective, common language for their exchanges. For millions served by the U.S. healthcare system, data can do the same.

Connecting the technology patients are using outside of the healthcare setting, such as fitness trackers and blood glucose meters, to the technology physicians are using in the provision of care, such as electronic health records and care management platforms, is one of the most effective ways to integrate people with their healthcare provider. As patient and provider utilization of this data increases, the healthcare system will be able to serve more patients more efficiently and effectively. And, in 2018, the ability to make care more efficient and effective is critical.

Studies show that over the next 12 years, the healthcare system will be short anywhere from 42,600 to 121,300 physicians [4]. Dissimilarly, both the number of patients in need of care and the complexity of conditions is expected to rise as a result of a growing obesity epidemic in the United States. By 2030, it is estimated that 42 percent of the U.S. population will be obese, leading to a steep increase in chronic conditions. This substantial rise in obesity indicates there will be nearly 8 million new cases of diabetes a year and 7 million more cases of coronary heart disease and stroke annually [5].

The numbers make it clear: Without the aid of technology, we will simply be unable to support the number of patients needing care in the very near future. The only way to deliver a higher quantity and quality of output—with fewer human resources—is through the deployment of technology. By using technology strategically and by implementing it invisibly, we have an opportunity to augment the care that clinicians and care teams provide to patients today to make it more efficient and more streamlined, in order to successfully manage the health issues patients will be facing.

Rather than reacting solely to patient calls and office visits, we can leverage the information and the data already available to us through existing technologies to more proactively respond to patient needs and intervene to encourage better health behaviors, with the aim of preventing the exacerbation of such conditions.

With effective integration between patients' health apps, wearables, and devices, and providers' EHRs and other clinical systems, clinicians can integrate the information that is already available in disparate sources into the healthcare system. They can seamlessly connect with patients via their device or platform of choice in a way that makes care more efficient. They will have the opportunity to combine remote data with the information collected during visits to garner deeper insights, advance clinical knowledge at the point of care, and make more meaningful, more proactive care decisions. In this way, we can make the technology and experience invisible to the patient, and we can make the experience invisible for the provider as well.

For years, we, as an industry, have been focused on innovation, on the creation of new tools, new data sources, and new solutions to solve problems and deliver care. The tools we need to reach these goals already exist. Patients are already generating health data that enable care teams to glean valuable insights. Electronic health records are already providing a platform in which providers and patients can access health history and other information.

We must now turn our focus to the integration of these innovative tools. And, we must think more critically about how we bring together technologies and services to design a healthcare system that delivers world-class patient outcomes despite a declining clinical workforce.

There will always be new sensors, new analytics, and new systems that come to market. But, we have the technology today to impact care and improve the quality of life for millions. Progress will remain sluggish until we bring today's technologies together and design for integration. We must design workflows and systems that integrate these tools in a way that is user-friendly and, ultimately, invisible, for all users—both the patient and the provider.

The ultimate healthcare experience will be realized when we can simply live our lives, healthily, and care is invisible.

References

1. "Mobile Fact Sheet." *Pew Research Center*. Science & Tech, 2018 Feb 5. [Cited 2018 June 30]. www.pewinternet.org/fact-sheet/mobile/.
2. Perez, Sarah. "U.S. Consumers Now Spend 5 Hours Per Day on Mobile Devices." *TechCrunch*, 2017 Mar 3. [Cited 2018 June 30]. Techcrunch.

com/2017/03/03/u-s-consumers-now-spend-5-hours-per-day-on-mobile-
devices/.

3. Schneider, Eric C, et al. "Mirror, Mirror 2017: International Comparison Reflects Flaws and Opportunities for Better U.S. Health Care." *The Commonwealth Fund*, 2017. [Cited 2018 June 30]. interactives.commonwealthfund.org/2017/july/mirror-mirror/.
4. Dall, Tim, et al. "2018 Update: The Complexities of Physician Supply and Demand: Projections from 2016 to 2030." *Association of American Medical Colleges*, 2018. [Cited 2018 June 30]. Aamc-black.global.ssl.fastly.net/production/media/filer_public/bc/a9/bca9725e-3507–4e35–87e3-d71a68717d06/aamc_2018_workforce_projections_update_april_11_2018.pdf.
5. Egley, Sharon. "Fat and Getting Fatter: U.S. Obesity Rates to Soar by 2030." *Reuters*, 2012 Sept 18. [Cited 2018 June 30]. www.reuters.com/article/us-obesity-us-idUSBRE88H0RA20120918.

* * *

Establishing a Culture of Innovation—Eliminating Barriers

Rachael Britt-McGraw

When I came into the healthcare IT field, there were some significant differences between the culture I saw before me and the cultures I had experienced in other industries up until that time. How healthcare defines innovation, from whom it expects innovation, and the acceptance of and valuation of innovation all differed. Whereas in some industries IT is fully expected to constantly innovate and continually improve business processes and profitability, this did not seem to be the case in the healthcare space. In fact, it almost seemed there was a mistrust of IT and a general mistrust of why IT would put forward new potential ways for operations to accomplish things. I had a lot to learn and determined that I would earn the trust of operational leaders and staff through transparency and dependency as I endeavored to learn and add value. Feeling that the best IT is always in lockstep with operations, regardless of industry, I met with the COO to learn his perspective on how we in IT could serve him. I will never forget his response. He said, "You just do the tech," and he turned and walked away. This reaction to an offer of service was completely foreign to me. I had never been so summarily dismissed by my customer, and I knew then that things were culturally quite different than in my previous roles, and I was not going to succeed at all if I "just did the tech."

In my opinion, good IT work is not fully attainable within a tech-only bubble. I had many years of very strong IT experience and education, having built top-shelf IT teams in many different industries and in many geographic locations. But I knew I would first need to learn all about the people, the business, the challenges it faced, and historically how it had overcome barriers before I could bring that experience and skillset to bear. I stood before a group of dedicated, veteran healthcare staff as a complete novice, not even familiar with the acronyms used in the industry. Their IT infrastructure was in dire need of overhaul, the IT team was in shambles and the morale very low, and they were facing installation, implementation, and adoption of an electronic health record system on a necessarily abbreviated timeline with no previous IT guidance or leadership. It seemed an overwhelming and insurmountable task. Looking back, that day was the beginning of establishing a culture of innovation in that I put forth no pretense—I communicated clearly my ignorance of the industry to which many of them had devoted their lives. I asked them all to help me learn their world, and I began to listen actively. They were surprised by the open communication and genuine desire to learn from them and how much value I placed upon their collective knowledge. And they gave me at least some idea why there was a lack of trust in IT.

Some shared that they had experienced IT staff members in the past who acted superior and talked down to them. They had never had anyone in IT that actually communicated clearly using non-technical terms that people without a lot of technical knowledge could understand and apply to their work environment and needs. They had seen arrogant IT people who were long on promises and short on delivery and who didn't seem to own those failures but pointed back at them as the problem. I explained that I would build an IT team that was service-oriented and who would under-promise and over-deliver as central parts of the IT team's culture. They may not have completely believed that would happen that day, but they would certainly support me while I tried. Little did I know, I was removing a barrier even then by simply being vulnerable and communicating openly with these fantastic people. And they responded warmly and soon communicated to the CEO and COO that I was a valuable asset to them. Since this was a trusted group of long-term employees, this endorsement helped remove negative perceptions of IT at the top of the organization chart as well.

We all worked closely together over the next two years, pulling long hours, correcting infrastructure, and implementing the electronic health record system and electronic practice management. By the end of that time, we had forged genuine and lasting trust, and we had all learned a lot from each other. Every step of the way, I communicated with our physicians and other providers of care so that everyone in the organization knew what was currently underway, its status, and if we were hitting or missing targeted deadlines and budgetary guidelines. The result of openly communicating deadline misses and failures and then eating humble pie while getting those projects teed back up again until they were successful was fertile ground upon which to turn out innovative ideas. The trust forged allowed operational management to feel comfortable that IT had their best interests at heart, and would be honest. Thus they were able to view potential innovations with an open mind and add in their ideas and suggestions freely. We implemented a patient portal and then an electronic means to exchange health records and new clinical communications systems. The sky became the limit.

After that, our next innovation was to leverage technology in the exam rooms to give those doctors who wished to use it more ready access to images, studies, and medical records when speaking with their patients privately in the exam rooms. This had been widely opposed before. From there, we have pushed deeper into innovative ideas, introducing proximity cards and strict security protocols. Today, we are continuing to constantly innovate and reshape the practice through partnering with a local, innovative entrepreneur who is developing new systems for patient intake, verification of insurance, and payment collection. We have other potential partnerships in the wings, and the excitement of innovation is a part of our environment now. I feel a great sense of accomplishment now as IT has a seat at the strategy table and is now expected to suggest innovations and keep the practice out front from a technology perspective.

In my experience, IT can be seen by the business in one of three ways: As a utility, an advisor, or a partner. As a utility, the business sees IT like the power company—flip the switch and the lights need to come on. These businesses see IT as only an "expensive overhead." If seen as an advisor, the business will reach out to IT for potential ideas when it faces a particular challenge or finds itself in a bind. These businesses see IT as a "one-off solution provider." But the real value of IT in business today is unleashed when the business sees IT as a partner. These businesses include IT concerns and potentials in business and growth strategy and weave in technology and the ability to continue to innovate and progress within its routine decisions. We've moved through these roles, and what allowed us to do so was open communication to remove existing barriers and produce trust.

* * *

Security as a Driver of Innovation

Mitchell Parker

Innovation is about empowering who you surround yourself with around a noble mission. A mission where the environment is constantly changing and over which you have no control. Where your focus must continuously change and strategies adapt. Where, despite the barrage of change, your motivation remains precise on excellence and quality. Innovation is forged from such circumstances.

It's not about trying to relate to frameworks or emulating people who are on the cover of magazines. It's about people. It's about a message of team-work and continually supporting who you surround yourself with to change for the better and to challenge yourself to do better not only as a team member but as a parent, family member, community member, and citizen. As a leader, it means that you have to continually set the example and put others before yourself.

The final product is one small, miniscule piece, and in the end, the product doesn't make a difference. It's the results of how you affect those you surround yourself with that matter. No one is going to remember you for a project you did. They will remember you for who you are and what you did (or didn't do) to improve those around you.

This essay is about making innovation real and practical. We're going to get there. Information security is the last item people think of when they think of innovation, especially in healthcare, where organizations are known for glacial change and a lack of it. Combine that with the fear, uncertainty, and doubt caused by misunderstandings of HIPAA and cybersecurity, siloed corporate structures, and fierce competition that stifles cross-organizational cooperation, and you have a situation that many people do not want to deal with.

We make it real and practical by consistently communicating our mission and values to our fellow team members. We are not about scaring people into conformance or being the HIPAA Police. We genuinely want our team members to do better every day and improve as participants in the shared mission of improving the health of our patients and community. You don't do that by focusing on the technology alone or being of the mindset that some product or service is going to improve statistics and make you look better in front of a C-suite or board that judges you like Damocles.

You do it by putting yourself out there every day and focusing on what your team members need to be able to improve not only the health of their patients and community but also themselves. These are the techniques we use to do so.

The first one is customer service. We emphasize empathy, listening to our customers, and working with them to guide them through complex processes and methods they do not understand. We look at everyone as if they were our parents, friends, or family. We take the time to explain what we do. The news media scares people about information security. We combat that fear with understanding, calm, and support. We deal with people at genuinely bad times in their lives, and we understand that. We want them to feel better after calling us and that they've contributed to improving a situation, not making it worse. Bad customer service earns you a reputation and does none of that. It makes people less likely to call you or more likely to avoid you when something does happen.

We have two sayings we live by. The first is that "we are nothing without our customers." The second is that "we want people to trust us enough to call us when they see something." If we do not value our customers, then we are not trusted enough to work as part of the team.

Second is understanding. We approach resolving issues by continually looking at how we can address the real issues instead of putting a bandage over a gaping wound. We take the time to speak with our customers and understand their concerns and issues and focus on what the real issues are. We want to solve the business problem, not claim that some new "innovative" technology is going to fix it. I would rather have my team or a vendor take longer to understand an issue, architect a solution, and get a practical date to implement it than attempt to put something together that looks good on paper that will probably fail given unreasonable constraints.

The major issue we see with cybersecurity that we have to deal with is that products are continually marketed as being innovative and able to solve issues. From 15 years of being in this field, 10 as a CISO across two organizations, I have learned that is not the case. We have observed that products by themselves ultimately do nothing without people or processes they can follow. The major challenge is that every day the CEO on downward are presented with solutions or products presented as innovations, and oftentimes we are asked to vet solutions or take phone calls because someone targeted our C-suite as a way to get to either information security or infrastructure. These people don't understand our business or our real risks, and many think they can use fear to scare our executives into acting.

When we interview people for risk assessments or risk management plans, we do so with a mindset of being able to understand what the risks are, how to reasonably and appropriately address them, and how to demonstrate that on a continual basis. Technology is a minor piece of this. A major focus of what we do is to develop that understanding across the entire team, not just a targeted few.

We let people know that we care about them and their success and that we want to address these risks because they will help the organization better fulfill its mission. We don't want to have people drop everything and act out of fear because you can only do that for so long before alarm fatigue sets in and people ignore you. We want to do this on reasonable and yet aggressive timeframes. No one is going to give you the time of day if you don't understand them, their needs, or their concerns. They're not going to be empowered to change if you don't show confidence in them or the mission. If you don't approach security as a team, you will lose.

If you cannot demonstrate a change so that others understand that it will improve the ability of the team to deliver on the mission, then it's not worth doing.

The third is continuity and continually putting yourself out there. You're not always going to succeed. You will have vulnerabilities. You will have data breaches, malware, or successful attacks. You will fail. The day of the project is over. There is no longer a beginning, middle, or end. There is only a cycle of assessing, planning, addressing, and repeating that ad infinitum. You will not be perfect. However, you will always be there for the customers.

You need to communicate that message continually. We deliver messaging on a regular basis to the world, leadership, the team, and to our own team members in IS and InfoSec. People know us and remember us from the messaging more than the team members. We write articles, present at conferences, attend conferences, and speak to our internal and external customers regularly because we have to deliver the message that we need to think about lifecycles, not just about a beginning, middle, and end.

We need to think about the impact any change has on the organization, on its people, and on its processes and make sure that we communicate them out. The best example we can give is when we rolled out our anti-phishing training, part of which involves sending fake phishing emails to our users. We held off on this until we had a communication plan that involved upper management, the service desk, the email team, and the infrastructure team because we wanted to meet the objectives of good customer service, having people reach out to us, and preventing the issue of having people complain to senior leadership and being pressured to stop the campaign. We worked with our public relations team to craft not only the phishing messages but also a one-page document that was sent to senior leadership that emphasized the connection to the mission of the organization, reducing risk, and preparing to address real phishing emails.

What has happened since is that our senior leadership team has become some of the most enthusiastic supporters of the program and regularly send in the phishing messages we send out. We have received very few complaints this time. Most important is that we involved leadership through good communication, which was a force multiplier.

We regularly prepare internal presentations for our team. Every month we prepare one for our Privacy and Security Committee and IS teams that specifically focuses on security events and breaches that occur outside the environment in a non-technical manner. As part of these monthly presentations, we do mini risk analyses on the events and what we've found. We focus on two key items, which are takeaways for our audience and internal changes we make based on the findings.

No matter how hard it is, or how much you don't want to, you have to put yourself out there every day with consistent messaging. It's not easy to deliver a message consistently and make sure that every communication articulates like an accounting statement. However, the message is remembered longer than the person who delivers it, and part of that message is that changes and innovation are continual. As a leader, you need to set that example. Your team will follow what you do, whether it is conscious or not. Innovation comes from setting the example and following it yourself.

The fourth is self-review. You are not perfect. You never will be. The more you attempt to present that picture that you are right 100 percent of the time, the more you will ignore the opportunities and chances presented to improve, innovate, and create real change that matters.

The purpose of a risk assessment is not to check a box for the Office for Civil Rights (OCR). It is an opportunity to objectively look at your organization and where you can change for the better. It's not a scoring exercise to see how good you can do and hopefully look good for OCR. It's about reviewing what works and what can change.

If you are not honest with yourself about what the risks are and where they are, you will not know where the real opportunities to innovate lie.

What we do to further reinforce this are two quantitative risk exercises. The first is the bi-annual risk assessment we complete for the risk and insurance team. This involves gathering and scoring risks from across the IS department that involve people, processes, and technologies. These risks also cover business and environmental factors. We use a scoring system designed by our risk team to measure them and submit the combined list to them twice a year.

The other exercise is the HIPAA Information Security Risk Assessment that we complete yearly. We made the conscious decision to use the same risk scoring system for both exercises. This allows us to score and prioritize our risks so that we know what top ones to address and, more importantly, why. It also gives consistent messaging to our leadership as to how we view it. When you have a limited amount of time to present, consistency is key to getting the message across quickly.

The fifth is change through innovation. According to *Webster's Dictionary*, innovation is the introduction of something new. There are two issues with innovation that can cause it to get stifled in favor of old familiar ways.

The first is not planning for it. The risk management plan, a traditional information security tool, uses the risk assessment to determine the path and plan forward for addressing risks in the organization. We use this as an opportunity, more so than the risk assessment, for interviewing our team and determining the best ways to address identified high risks. From our experience, we understand that most information security risks have their root causes in business process issues, not just technology. Part of what we do is to take more time to develop plans to address risks along a longer timeframe. We do this because we understand that a change for our customers' needs to be effective, well-communicated, long-lasting, and well-documented. We are asking people to change their processes to address a risk that they may not see as applicable.

We use risk management plans as an opportunity to connect with the rest of the organization and get their input and views on what we are doing and why. We want everyone to own the plan and work with us to implement it. Security cannot operate in a silo, and neither can our customers. True innovation happens through getting multiple parties working together on a measurable plan to address risks as an organization. We make sure to put it in writing, define metrics for success, and most importantly, develop a communication plan with four targets, which are the board, executive leadership/upper management, line management, and everyone else. We want people to know what we are doing, why we are working together on this, and how we define and measure success. We also want people to know that we are collaborating for the benefit of the organization and that measurable and successful solutions take time and effort and don't happen overnight.

The second is through innovating in silos. In many graduate programs, including the ones I attended, they teach that innovation is normally isolated to separate organizations still operating in the corporate structure, much like the initial internationalization attempts that companies undergo. While we don't expect companies to end up as a fully matrixed organization like Nestle overnight, we need to set two expectations in our organizations.

We need to encourage collaboration across departments, divisions, entities, and even competing organizations. In the world of information security, we are all facing common threats and challenges. With the limited resources we have due to low profit margins and the complexity of our environments, duplication of resources and lack of collaboration puts all of our organizations at risk. We openly collaborate with other institutions and use our memberships in HIMSS, Scottsdale Institute, and numerous other organizations as vehicles for working with our peers. If there isn't an event that addresses what we need, we will host it ourselves or bring people in to discuss it at one of several forums we use. We will present on what we do at conferences because we want to collaborate, innovate, and franchise our processes so that we set expectations with our peers and vendors. We also want to educate the security professionals coming up through the ranks now.

We put ourselves out there internally and externally with all team members, especially our medical staff, IS department, collaborative groups, and leadership. We want to work with everyone and volunteer to do so in the hope that others will follow. We make sure our messaging is on point with the audience and speak to their issues so that we can get everyone understanding and working together as a team to address our risks.

A major component of collaboration and innovation that often gets forgotten is that you need a good open-door policy. While this lately has gotten press due to its apparent use by a certain business icon to flag people "breaking rank," we have found that this use is not an isolated practice. If you're going to change for the better, you need to solicit input and listen to everyone, and if you use this policy to flag people, stop right now. This means that you have to make sure your organization respects and listens to their fellow team members.

One of the best groups of people for soliciting security changes has been desktop support. Traditionally in a lot of organizations that I have worked in, they have been treated as "grunt work," that is, easily replaceable and just "techs." We consider those words dehumanizing and disrespectful. They have found more issues and reported them to us than most other teams within the IS department and innovate every day out of need. There are only so many support team members and a torrent of tickets. They figure out more ways to be efficient and complete their job well. We recognize that and have several former desktop support specialists working in information security because of it. We want people with that mindset helping reduce risk in our organization.

We have also seen multiple teams in IS departments ignore them because they are "just techs" that do what they are told. This is as far from the truth as you can get.

Finally, you need to be constantly reinforcing the benefits of change and innovation. To truly support change in your organization, you have to show measured success. Innovation is good, but you need to have facts and statistics to show that the organization is improving. You have to consistently communicate with everyone, and you have to follow up. Part of evolving a team to innovate as part of its DNA is to make sure that it replicates and persists. The NIST Risk Management Framework specifies that you need to constantly follow up. Innovation cannot come in dribs and drabs. Change is constant. You have a choice to make, which is whether or not to harness it.

Innovation doesn't come from products, EMRs, or tools you buy. It comes from within, and it can take many forms. You have to empower the organization on multiple levels to be successful. It's easy to say that you do it in areas, but to really innovate, you need to focus around a mission and continually reinforce it.

* * *

Cultural Considerations for Innovation in Technology

Rosie Sanchez

Over the past 14 years, I have had the opportunity to work in various roles at a large community healthcare organization on the Mexican border, primarily targeted to an unserved patient population. These included roles such as a clinic front desk support staff, administrative assistant in the business office, support staff in the medical education department, and as an entry-level analyst in various IT department roles; ultimately leading me to my current role as Senior Director of IT/clinical information systems. Through my experience I have found that my passion lies in healthcare information technology. I learned that many of the manual processes I used in the early years of my career were leading to human errors that could have been improved with the use of technology. I also learned how process changes and workflow modifications (even those thought to be an improvement for the end uses), through the adoption of technology, could ultimately have drawbacks for both the frontline staff and patients. As a Hispanic female, mother, and daughter of elderly parents who are patients in this healthcare organization, I can directly relate to our patients' needs. I believe that innovative technology, when considering the needs of the population served, along with the people serving them, can open doors that will ultimately help organizations better support our socially and culturally diverse communities.

In order to be able to successfully implement a new technology, it's important to consider the approach used and the audience that will engage with the technology. Failing to do this can result in a project that either fails or has long-term growing pains or adoption issues. There is also the risk of ongoing issues related to the project, which can take a lot of time and resources away from both the implementation team and the clinics expected to use it. Understanding your audience/target group is extremely important. Too often, those at the top making high-level decisions cannot accurately gauge the impact their decisions can have on clinical staff and patients. To be successful, it's my opinion that the best approach is to shadow the clinical staff and their interactions with the patient population if you do not already understand the details of their workflow and how they interact with patients. Speaking with the staff, observing them, and understanding how the patients use healthcare and interact with any existing technology is vital. I strongly recommend this approach to any healthcare IT leader. Immerse yourself in their world, and you will be able to find ways to successfully implement innovative technologies in even the most seemingly difficult populations.

In our healthcare organization, as mentioned before, we have greater than 80% of the population as Hispanic. This is composed of people that have lived on the border their whole lives, the majority of which speak Spanish primarily at home, are ESL, speak no English at all, or are first generation. This applies to both the patients and the frontline staff. A population like this has very different cultural needs and considerations. Many also may be older or unfamiliar with technology. As a result, historically, we have had very low adoption rates of technology in the past. To overcome this, we have had to identify potential areas of strength in this population of people, as well as come up with work-around solutions for weaknesses.

Our last big implementation was of an automated patient intake platform. This was supposed to reduce clerical/human errors at the front desk and improve the speed at which patients were registered/checked-in for appointments. Many thought that due to the use of technology, patients would reject or decline to use the platform. This could have been an obstacle that prevented successful adoption. However, even though the patients themselves may have been a largely older and Spanish-speaking group, in Hispanic cultures, it is common for adult children and grandchildren to accompany aging parents/grandparents to doctors' appointments. Knowing this, we decided a better way to target the patients was to use this to our advantage and target the family members. Our patients trust their family members and rely upon them so much that getting the family member to accept the technology (when many of them are younger and heavy users of technology outside of the healthcare setting) was a very successful approach in getting the patient to accept it as well. Even our organization's leadership was surprised at the quick and successful adoption of the patient intake platform.

We also used this same approach to improve our enrollment/usage of the patient portal. Since many of our patients do not have email addresses themselves or access to a computer, we took the alternative approach to having the front desk target the family members to enroll on their patients' behalf. By explaining to the patient or caregiver the benefits of having an electronic location to store pertinent health data that could be accessed by their family member for current information, or possibly shared with other healthcare providers in the community, we were able to almost double our percentage of enrolled patients from one year to the next. The front desk staff saw the benefits of enrollment and were able to sell the patients and their family members on enrolling in and using the portal with much more success.

Those were just two examples. Putting this concept into action in any organization is feasible. You have to first consider the needs of your patients or staff. Look at the unique characteristics of the cultures and values of the population you employ or serve. It's important to try to view the differences or what appears to be a problem as an obstacle that needs to be overcome. If you yourself do not personally have experience with the population you're assisting, it's important to consult with people that do. They can provide personal insight into the potential issues that may be preventing you from success. We took a problem of non-English speaking patients who typically do not even own their own smartphones and found a way to have unique aspects of their situation benefit them: Their propensity to have younger, technologically proficient family members heavily involved in all aspects of their lives, especially healthcare. I personally have experience in this area and could really relate to the patients and how they accessed care, but many of the other team members that worked alongside or in leadership, who approved the projects or created the plans, did not. My input (along with others) helped to convince them that successful implementation of innovative technology was possible here with the right approach.

Another way we were able to get personal feedback from our diverse group of patients to determine their attitudes toward a potential new technology venture (telemedicine) was to conduct a survey. The Neurology Clinic was considering the possibility of bringing in telemedicine for patients with epilepsy that, historically, had high rates of cancellations due to transportation issues, families not being able to accompany them to appointments, etc. Many stakeholders and leadership members doubted that our patients would be interested in using this technology based on their backgrounds, speaking primarily Spanish, their ages, or the assumption they had limited technology access. To really find out what the patients' views were, we decided to create a simple paper survey, with English on one side, Spanish on the other, and have the front desk staff in the clinic hand it out to all patients while they were waiting. Due to the fact that it was given on paper, it was easy to administer and required very little use of resources and no new technology. Also, because both languages were presented, no one had to spend time asking patients which language they preferred to be asked the questions in. From the survey, we discovered our patients were really interested in the new technology, and even those without access personally to technology said they would be able to access through family members. Another surprising result of surveying the patients was that their interest wasn't correlated to their preferred language. Additionally, many were very enthusiastic and wrote comments requesting to bring the technology soon.

When having meetings with internal employees regarding new implementations or potential tools, be sure to involve the front staff, or create committees/groups that allow them to feel part of the process as well as voice any concerns they may have. These are the staff that have to really believe in the tools you are trying to introduce and understand how to use them effectively so they can also help assist patients. We did this when creating the telemedicine survey for our neurology clinic, and the staff really was dedicated to passing the survey out to each patient. You will have a much more successful implementation if staff concerns and needs are addressed beforehand. The frontline staff are a key component to this since they interact with the patient first. So it's also important that they are on board and really understand how any new tool will improve the patient experience or make their jobs easier.

The importance of knowing the culture you are supporting is paramount to success in innovation. We have a unique opportunity in my border region to serve a special population of underserved patients. In the future, given the way that healthcare is moving, and many initiatives by CMS and other government-related agencies, it appears that engaging patients with innovative technology tools will only increase in importance. This is just one aspect of healthcare, but it's important because patient engagement with technology affects things on so many different levels. Even meeting requirements of programs, such as Meaningful Use or MIPS, or implementing technology to streamline co-payment processes, or enrolling patients into an HIE, require the organization to successfully convince their patients to be engaged with the tools. As technology continues to evolve, so does the need to understand the culture and environment of your unique healthcare organization.

Chapter 6

Stress Simplicity

> *Do not overcomplicate a solution to a problem; keep the following*
> *principle in mind: "When you have two competing theories that make*
> *exactly the same predictions, the simpler one" is better to implement.*

It seems counterintuitive, but the majority of innovations are rather simple.
The temptation to take a problem and create a complex solution exists in
most of us. We tend to overthink an opportunity and therefore overengineer
a fix. Innovation is often as basic as developing an elegant yet simple solution to a complex issue or opportunity, not the opposite way around. If the
innovation can't be easily explained, start again.

Advancing Innovation

Daniel Barchi

Simply begin. Most innovation fails not in its execution but at its inception.
Health systems are built and tended by bright people with innovative ideas, yet
many of their ideas are never put into action. If perfect is the enemy of good,
over-planning is the enemy of starting. When facing a monumental project, the
planning is almost always easier than doing the work, and more planning can
feel safer than starting a project and facing the risk of failure. Yet time spent creating plans and budgets that foresee every contingency is time that is not spent
learning lessons by putting ideas into play. General Colin Powell famously said,
"Once you have information in the 40%–70% range, go with your gut." I wanted
to create an environment where our team felt loyalty to the project, not the plan.

DOI: 10.4324/9781003372608-6

This "learn-as-we-go" approach has far-reaching benefits. I find that it offers a subtle change in mindset. Instead of focusing on a plan, our eyes are open to solutions. By simply beginning, we are freed of the need to defend a plan, and we are better positioned to receive input from our users who are normally the people on the front line of medicine. All too often, the IT team is viewed as the reason why nurses cannot log in or pharmacy labels do not print, or worse, as remote people who push new systems upon an already pressed staff. When we simply begin, we are automatically part of the same team, and anyone in healthcare knows teamwork is an essential part of delivering the best patient care.

As our IT team prepared to implement a refrigerator temperature monitoring system, I thought of Powell's sentiment and my own experiences. Years ago, the responsibilities of a hospital IT team were generally limited to supporting mainframe and desktop computing. As technology has developed, even hospital beds are connected to the network. Biomedical engineering has aligned with IT, and we are now responsible not only for computers, printers, and televisions, but IV pumps, MRIs, and refrigerators that store vaccines, breast milk, and medicines.

To ensure that these critical medical supplies are kept between 1 and 3 degrees Celsius, nurses on hospital floors used to manually check the temperatures twice a day and log the results on a piece of paper attached to each refrigerator. Although such monitoring seems a simple task, nurses are busy, and hospital safety inspections routinely identify fluctuating refrigerator temperatures as a major problem. With the goals of freeing nurses of this responsibility and achieving reliable daily monitoring, our team installed temperature probes in each refrigerator and networked them back to a central monitoring station. No one we knew had done this before, so our team struggled with the technology, clinical workflow, and the process for addressing temperatures that fell out of the parameters for safe storage. I met with the team twice in January of 2016 as they bounced back and forth between hospitals, the data center, clinical meetings, and planning sessions. They wanted to get everything perfect for an April 1 "go-live" across six hospitals, but I was skeptical that we knew enough to count on a single, flawless launch. "Just turn it on now in January for one nursing unit and see what you get," I told them. They were reluctant because they had not yet resolved problems with network communications and nurses' notification. They were also concerned that refrigerators would alarm at the central station every time the door opened for a nurse to retrieve a vaccine. "Let's just start," I insisted, "and see what we get."

So, we simply began and turned on remote monitoring for four refrigerators. One day later, we were glad we did. We learned very quickly that when a nurse checks the temperature at the same time every day, we got the temperature we expected:

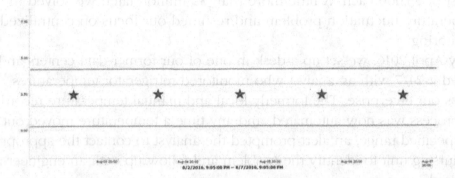

When we used IT to remotely monitor and sample the temperature every five minutes, however, we found that the temperature fluctuates wildly in between readings:

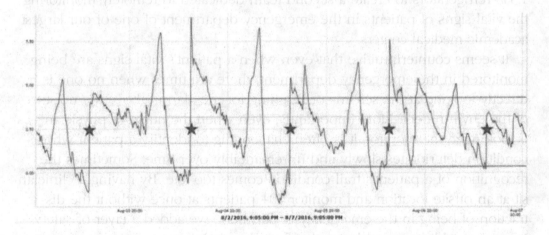

In the midst of trying to perfect the system, we did not consider the variable at the heart of this innovation—the temperature of the refrigerators themselves. Our eyes were opened. As it turns out, the refrigerators we had in place were not meeting our temperature specifications even when the refrigerator doors remained closed. Our computer-based monitoring revealed that each time the compressor started or stopped, the temperature dropped precipitously or spiked above the threshold.

When we began the project, our focus was reducing the time nurses spent on equipment monitoring so they could spend more time with patients. By starting our work rapidly, we gained key insights early and recognized that we needed to solve a different problem entirely. Of the 1,350 refrigerators humming throughout our six hospital campuses, 780 had this issue and needed to be replaced with upgraded refrigerators at a cost of $3,000 each. A little more than $2 million later, we solved the temperature fluctuation problem and resumed our focus on centralized monitoring.

By April 2016, we set up a desk in one of our former data centers and staffed it 24/7 with an analyst who monitored refrigerator temperatures across our enterprise. The formerly local and manual temperature recording process was now automated, and any time a temperature moved out of the specified range, an alert prompted the analyst to contact the appropriate nursing unit to identify the problem and follow up with an engineer as required.

While this was clearly a step forward, we realized that centralizing alarms and data feeds presented additional opportunities. Now it was time to get creative. We took advantage of the space and the 24/7 staffing inspired by 1,350 refrigerators to create a second team dedicated to remotely monitoring the vital signs of patients in the emergency department of one of our largest academic medical centers.

It seems counterintuitive that even when a patient's vital signs are being monitored in the emergency department, there are times when no one is directly looking at the screens or listening for the alarms above the noise of the environment. More importantly, even when the nurse or physician is looking at the monitor, it is often challenging to identify a patient whose condition deteriorates slowly and unremarkably over time. Sometimes the recognition of a patient's frail condition comes too late. By having a clinician sit at an offsite location and monitor 50+ patients at once without the distraction of being in the emergency department, we added a layer of safety. Instead of adding to the demands of an already busy environment, we now had an experienced clinician whose only responsibility was monitoring the data of each patient. When the remote clinician identified a significant change in the condition of a patient, the patient's assigned nurse was immediately alerted to the situation. Within the first few weeks of remote monitoring, we had more than six of what we called "great catches" in which the patient's deterioration was identified early and in time.

Instead of building an entirely new remote patient monitoring center, we had taken the former data center, added minimal equipment and staffing, and created what we called our Clinical Operations Center, or CLOC.

Other innovations followed—we found that in our health system, new ideas sometimes would not be put into practice because of a lack of space or resources to make them happen. So, we began to expand our capabilities at the CLOC.

We next moved operators from four of our hospitals into the CLOC and sought ways to maximize their productivity. Formerly, they had simply answered phones and transferred calls. When a patient called for a prescription refill, our operators queried the patient for data, completed a form, then faxed it to the appropriate physician's office. We saw that these talented people were underutilized and could help us improve and streamline the process. We taught them to use our electronic medical record, trained them to coordinate telemedicine video calls, then promoted all of them from the role of call center agents to account representatives with an average 13% increase in pay. In their new roles, they became active participants in the clinical care process, increased first-call resolution by solving issues themselves, and even reduced the time patients waited for prescription refills from an average of eight hours to only five minutes. Patient care improved, and our staff advanced their training. Our innovation in this case was simply to take the people and resources we had and make them more efficient by focusing on their training and their workflow. IT teamed up with patient experience, and we were all better for it.

We took this simple approach to innovation to new heights in the summer of 2016 when our Chair of Emergency Medicine said he would like to try telemedicine visits within the busy emergency department of one of our hospitals. While urgent care visits by video are nothing new, he thought we could make physical visits to the ED more efficient by offering patients with less serious conditions the option of a virtual visit with a physician. Like the CLOC, we decided to simply begin.

Within two weeks, we installed video equipment in an existing room in our emergency department (ED), added a camera to an existing physician office 200 yards away, and started seeing patients in what we call NYP OnDemand ED Telehealth Express Care. Patients are greeted at the front door and triaged by a nurse. Those whose injuries or illness are mild are presented the option of a traditional visit (about 2.5 hours) or a video visit (about 30 minutes). The patients who opt for a virtual visit enter a private room and immediately begin a one-on-one consultation with a physician by video. At the end of the visit, the remote physician sends discharge instructions to a printer in the patient room and can send prescriptions electronically to the pharmacy of the patient's choice, when necessary.

We simply began by leveraging people, space, and technology we already had on hand. By starting rapidly, we learned what worked and what needed improvement within two weeks of our initial idea. Patients of all ages loved the service—when a 21-year-old patient and an 89-year-old patient were seen back-to-back, it was hard to tell whose comments ("This was the coolest," and "I am going to tell all of my friends") were whose. We started with 10 patients, rapidly tweaked the process, saw 50 more patients, evaluated the process, then opened it up broadly. We have now seen more than 10,000 patients through the service and reduced ED wait times and revisits.

Net promoter score (NPS) is a way to gauge the satisfaction and engagement of customers. Ikea has an NPS of −9, Lego an NPS of 6, and Apple an NPS of 72. Our ED Telehealth Express Care Service is well-liked by our patients and has an NPS in the mid-90s.

This service is one of ten different telemedicine modalities we now offer patients. In each case, the offering was not the result of years or even months of planning but a rapid response to an identified need. When we realized that nursing homes would send patients to our ED in the middle of the night "to be safe," we created a TeleNursing Home service that allows nursing home staff to connect with our ED physicians 24/7 to determine if a patient really needs to be transferred. When we realized that patients in our EDs sometimes waited up to 24 hours to be seen by a psychiatrist or transferred to a facility with psychiatry coverage, we used existing equipment and our own fellows to create a TelePsych service that shortened that wait to within two hours.

Innovation is not consistent with multi-year plans. Rapid implementation and real-world testing quickly demonstrate what innovation has value. While clinical interventions that have a direct impact on patients can and should undergo rigorous clinical trials and testing, operational process improvements can be put into place rapidly and refined using real-world feedback. The mantra to fail fast can only be followed once work has started with a process in place that can succeed or fail. By simply beginning, we create an environment that is free from the stigma of mistakes. As hospitals strive to identify risk and prevent mistakes, it is important for IT to follow physician leadership and do the same. With loyalty to project before plan, a willingness to embrace rapid change can significantly improve not just operational efficiency but patient and employee experiences as well.

* * *

Stress Simplicity

Marc Probst

Consider gauze. Consider that loosely woven, cotton fabric in ways that no one does these days. Then consider how healthcare takes gauze for granted—how it's a staple in care provision but rarely marked for its importance. Now, consider that once in history gauze was innovative but outrageously simple.

Simplicity in innovation doesn't translate to simple ideas. Instead, simple innovation solves problems for people. Yes, there is much to be said about the glory of volcanic and disruptive innovations, but such innovations are few and far between and often unearth a bevy of new problems to be solved. Simplicity in innovation is almost harder to achieve; overcomplicating ideas is easy. Reinventing the wheel, as they say, is most often a delay in progress, and dreaming, as opposed to doing, isn't innovation.

We don't have to go back to ancient Palestine and the invention of gauze to see how simple innovations can have lasting impacts on people and their lives. We need only go back 15 to 20 years and enumerate the slew of simple innovations that have shaped—and continue to shape—healthcare, like new medication delivery systems, electronic health records, and automation. Innovations that improve care the most quickly are usually about the application and use of existing tools in new ways, just like when someone discovered that gauze doesn't adhere to wounds and tissues like denser fabrics do.

If you want a culture of innovation in your organization, I suggest simplicity. In my organization, Intermountain Healthcare, we've established a framework of innovation that touches every employee. We also make sure that all 37,000 of our caregivers are invited to the innovation party. No one is exempt from idea creation, and sometimes some roles are expected to generate a given number of ideas each year. No idea is too small even if the market would snub its non-complexity. We are rewarded through every contribution no matter how seemingly small or mundane. No idea is too simple to be explored as a potential innovation or innovative use of a tool.

I urge you to keep Peter Drucker in mind as you move through this essay:

> *An innovation, to be effective, has to be simple and it has to be focused.*
> *It should do only one thing, otherwise it confuses. If it is not simple, it*
> *won't work. Everything new runs into trouble; if complicated, it cannot*

be repaired or fixed. All effective innovations are breathtakingly simple. Indeed, the greatest praise an innovation can receive is for people to say: "This is obvious. Why didn't I think of it?"

Here's how I unpack Intermountain's success at simple innovation.

We Seek to Solve Problems for People

At the root of all innovation is need. That's why focusing innovation efforts on solving problems that affect people has a high return on investment (ROI). At Intermountain, we take multiple approaches to doing just that. Here are just a couple.

One initiative at Intermountain, something we call *Design for People*, seeks to observe people in the context of their daily work while looking for ways to make the work more efficient, safe, and complementary to the goals of healthcare. We've found that developing innovations that make a difference starts with a deep understanding of the work our physicians, clinicians, staff, and patients are doing. By understanding the context, we are best able to find simple innovations that improve the way the work is done.

Case in point, our *Design for People* team went to two of our hospitals to observe the work in the endoscopy labs. During those visits, the team noticed that the stationary position of sinks the Endo Techs used in cleaning and preparing scopes was of concern because of the various heights of Endo Techs. To remedy the situation, the team recommended the installation of variable height sinks, which resulted in less back strain for the techs and more efficient use of the sinks.

Such small changes can have big impacts, so we encourage all our caregivers to look around their work areas for ways to improve processes. We empower caregivers to take the responsibility to observe and share ideas for improvement, then we collect those ideas, assess them, and implement those that we can.

Collecting ideas is at the heart of innovation. As leaders we often think that we have all the information we need to address issues efficiently and effectively, but the truth is the information we have is usually lagging or outdated by the time we have it. When an issue is discovered in post-problem reports, it can be too late to institute a change that can really affect outcomes.

That's why Intermountain has invested time and resources into creating a series of tiered escalation huddles that begin on the very frontlines and ensure items that need executive level influence reach executives in a timely fashion.

Intermountain learned early on that it's insufficient for ideas to simply go from the frontline to manager; ideas sometimes need to be escalated even further up the chain of command. Our escalation huddles make that possible each morning as caregivers meet in Tier One huddles to discuss problems and ideas for improvement that, when warranted, reach the executive Tier Six huddle by 10:30 A.M.

We also make it simple for caregivers to submit ideas through an electronic tracking tool. Some caregivers are expected to submit a quota of ideas per year, and that quota is tied to performance evaluations and goal completion. Simple innovation, however, doesn't stop with caregivers today, with the rise in healthcare consumerism, we recognize that innovation must also embrace and take the burgeoning trend of healthcare consumerism into account.

Here's my note on consumerism: Smartphones and other smart devices are enabling consumers—patients—to monitor everything from blood glucose to heart rhythms. Sensors are becoming ever cheaper and being integrated into these devices. The tech is simply exploding, and gamified incentives, like badges, encourage consumers to use these tools. This unprecedented self-gathered data gives patients insight into their own health that, when leveraged correctly, will significantly change how consumers access, use, and select healthcare options. Simple innovation must embrace this trend and begin looking at how these consumer-facing tools can create efficiencies the like of which healthcare hasn't seen. But, the smart phone in healthcare won't stand the test of time if we don't start innovating around its footprint. Electronic medical records need to fit that mold and become as streamlined as apps like Uber's.

We've Built a Platform for Innovation

Beyond our *Design for People* initiative and escalation huddles, Intermountain has also created a program to take ideas that offer market potential—and the people who come up with them—to the marketplace.

The Intermountain Foundry, part of our business development area, provides structured support and resources to advance caregiver innovation. The Foundry uses a proprietary, market-tested curriculum to help internal inventors, innovators, and entrepreneurs determine the commercial viability of their concepts and scale them throughout our organization and the broader healthcare industry.

Through a structured eight-week course, innovators refine their concepts, validate market opportunity, identify methods of product development, and create a plan for growth, including strategies for taking a product to market and funding options.

Each year, Intermountain selects—from a competitive applicant pool—the four most promising innovators to present their innovations to the Intermountain Innovation Fund Steering Committee and make funding requests. The Foundry gives perspective to the meaningful impact that new innovation can have on healthcare delivery.

Healthcare delivery is the center of our Kem C. Gardner Healthcare Transformation Center, available to a worldwide audience of healthcare leaders and innovators—such as those in our clinical programs—who seek to improve healthcare through innovation. The center will house two of Intermountain's nationally and internationally recognized institutes—the Healthcare Delivery Institute and the Healthcare Leadership Institute.

We Collaborate with Others for Success

The collaboration that comes from training the best leaders in healthcare delivery and leadership is one way Intermountain partners with others to share the fruits of innovation. But, sometimes a problem is too big for one individual or organization to solve. That's why it's important in today's healthcare industry for like-minded organizations to band together to solve problems.

Intermountain has partnered this way in several areas, but most recently we've teamed up with other organizations to help patients by addressing shortages and high prices of life-saving medications. So far, 120 health organizations that represent about one-third of the nation have expressed interest in joining Civica Rx, a not-for-profit organization.

Civica Rx seeks to stabilize the supply of essential generic medications administered in hospitals, many of which have fallen into chronic shortage that places patients at risk.

Such partnerships make solving big problems more manageable and create a community within the healthcare industry that solidifies the aims and goals of medicine—to encourage, maintain, and restore health.

We Share Results for Posterity and the Greater Good

It's been said that innovation is a team sport. That's why Intermountain often shares its extensive clinical experience, the results of years of improvement data, and our commitment to transforming the healthcare industry. One way we do this is through our clinical programs and the critical research and development they participate in.

Since the late 1990s, Intermountain's clinical programs have been working to set clinical improvement goals and leading system efforts to accomplish those goals. However, we find that it's not enough to just develop those goals within the system, so we share the findings we reach with others to improve care nationally and internationally.

Today, our clinical programs identify, develop, and deploy best practice protocols and guide Intermountain in a multitude of processes and decisions including improvement goals, clinical staffing models, regulatory standards, purchasing, medical necessity requirements, pay-for-performance metrics, education, and documentation standards.

By sharing the results of this work—and the methods for conducting such work—Intermountain helps others to innovate, truly honoring the idea that innovation and healthcare are a team sport.

So, I urge you: Think simply, but create a culture of innovation in your organization. You can do this based on four simple areas: Solve problems for people, build an easy-access platform for idea generation and innovative thought, and collaborate with others. Healthcare needs every idea and innovation it can get right now. We're facing such tremendous change and stressors—like quality, diminishing resources, and clinician burnout—and we must syphon and harvest every idea if we want to continue helping people. Innovation will only prosper if we keep it simple and seek to make our solutions as common as gauze.

* * *

Healthcare Digital Transformation

Joey Meneses

Healthcare must continually evolve digital capabilities in line with changing business requirements. Through optimization and modernization of technologies, organizations will increase return on investments and enable business agility and growth. Here are three key drivers:

1. Healthcare provider and care delivery: Adopt and evolve digital technologies to optimize clinical workflows, improve care outcomes, exceed patient expectations, streamline business processes and maximize profitability.
2. Consumer engagement, value-based care and ecosystem collaboration: Adopt digital technologies and approaches that delight consumers,

improve healthcare quality and outcomes, coordinate healthcare services and decrease the total cost of care.

3. Mergers and acquisitions: Adopt digital capabilities to quickly consolidate independent ecosystems when it makes sense. Rapidly increasing demand and rising costs are forcing the healthcare industry to evolve from disconnected entities working independently to a seamless and efficient ecosystem acting as one.

Healthcare organizations must focus on executing the vision and strategy of their organization and understand how technology can enable that vision. They must communicate how technology can enable their organization's business model, operating model and the broader healthcare ecosystem in which they operate. The need to improve outcomes by exploiting information and technology to enable digital care delivery and empower clinicians and consumers is stronger than ever. At the same time, providers must capitalize on the power of digital technologies to improve operational efficiencies, remove duplication and optimize revenue cycle management.

Leadership is key as this is about transitioning from the existing model of operations to new ones. Often, this phase also demands rethinking the boundaries of the organization as it moves from traditional linear supply chains to ecosystems. It requires a high level of digital transformation maturity. The key to more successful digital transformation is to not skip ahead: Start with step one and invest the focus and resources to get it right. Growing your organization's digital maturity through the digital transformation corporate learning curve will increase your chances of success.

Digital transformation can be a response to changes in the market or competitive landscape, or it can be driven by a desire to improve efficiency and productivity. In many cases, it will involve a combination of both. Whatever the reason for embarking on a digital transformation journey, it is important. Digital transformation is about transformation that's driven by digital technology. In everyday life, this could be the rise of smartphones and apps. Think along the lines of Google Maps, Uber and search engines. For businesses, it's using digital technology to improve existing processes and systems or change the way the company operates, with the ultimate aim of boosting customer or shareholder value.

In other words, digital transformation can boost your value proposition. Just on the horizon, many business executives believe IT will create new value propositions by innovating and driving tech solutions that ultimately determine positive business results.

The first step executive leaders must take to design a digital business transformation is to redefine the enterprise value proposition and then address the dependencies between strategy, capabilities, financial, business and operating models. A new and differentiated value proposition is at the heart of a business model change. Changes to the value proposition cause ripple effects that create dependencies and business design adjustments across the business and operating models.

Adjust your enterprise's value proposition to stay competitive in the market by asking the following questions:

Is your industry going through disruptive forces today?

What is your digital ambition level? Are you aiming to optimize the existing business model and/or transforming/expanding the existing one into something new and additional?

When observing near-term industry disruption, are you prepared to pivot and adjust your value proposition?

While technology may lie at the heart of digital transformation, companies must recognize the fact that a relevant value proposition—and real transformations—should be driven by what consumers want.

These are true transformations because they challenge the existing processes, structures, and capabilities of the organization and require new ways of working.

Digital health connects and empowers people and populations to manage health and wellness, augmented by accessible and supportive provider teams working within flexible, integrated, interoperable, and digitally enabled care environments that strategically leverage digital tools, technologies and services to transform care delivery.

* * *

Simplicity—Three Guiding Principles for Payers

Sakshika Dhingra

Payers hold a unique and significant position in the healthcare ecosystem. Payers are uniquely positioned as the aggregators of information and data to use it to drive health value for our members. Payers are also the next-best-action coordinators in the members' care journeys.

Naturally, that makes it a hotbed of tremendous opportunity.

Here's a quick litmus test that I use anytime I am assessing where we as an industry are on the innovation continuum. And I call this "What I would want for my family test."

And the answer is, "We have a long way to go" simply because when it comes to our own family, we call in favors, ask our friends for help, and do whatever we can to skip the line. Yes, we do. And that's opportunity.

It is extremely important that payer organizations realize the importance of innovation within their space to propel the whole landscape toward a better more integrated future where care seekers are able to have a unified experience.

In this essay, I attempt to shine a bright light on three basic guiding principles that payer organizations can follow to create an atmosphere that is not only conducive to innovation but fosters it from within.

As simple as this first principle may sound, I was amazed at how often this gets forgotten in day-to-day business discussions—keeping consumers at the front and center of every initiative. Being able to create on-going value for our customers is what distinguishes the most successful organizations from others. I say "ongoing" because it's not a onetime thing and requires a mindset of excellence around the consumer. This would mean that the team puts the customer at the center of all their decisions, from their daily priorities to their processes to how they spend company dollars. Developing a mindset of how the customer will benefit from our actions will help connect decisions to serving the customer. It will also help connecting actions to a greater purpose while strengthening team and culture. It's a great framework and helps embed continuous innovation by improvement into the very fabric of an organization.

One of the biggest barriers in creating a consumer centric organization is that organizations are structured around the work they do within the four walls (virtual or otherwise). They start by organizing themselves by technology, then departments, and then there are some that are further along, and they organize themselves by consumer journey steps. But think about it from a consumer perspective. A consumer does NOT get value from these touchpoints; a consumer gets value from the whole stream. And frankly, most organizations don't understand value. The key is to understand our consumers—whether they are providers or members—and what they value from our service. Once this is done, orgs can immediately see how structuring by specialization created a labyrinth of an org where no one fully understands how value is created.

Organizing around value streams is the first and the most foundational step a payer organization can take toward becoming a truly innovative org. Bringing everyone who touches a stream closer and together will enable them to co-create solutions that would maximize the value for our consumers.

Now that we have the form right, let's talk function. Form follows function is a *principle of design* associated with late 19th- and early 20th-century architecture and industrial design in general, which states that the shape of a building or object should primarily relate to its intended function or purpose. And it's extremely relevant to my next guiding principle, that is, architect the right operating model.

An operating model *describes how an organization delivers value*. Healthcare is an industry that experiences constant change and disruption, and the digital acceleration during the pandemic has put us even more behind than other industries like retail as far as innovation is concerned. Naturally, market obsolescence is a growing threat for many organizations out there, especially with new more tech savvy market entrants that have the technology and the resources it takes to innovate. The result is a wave of transformative initiatives these organizations are going through. And many transformation efforts begin with reinventing the operating model.

The key to reinventing your operating model is to have a unified vision— a central strategy to which the model can be aligned to. Nothing brings an organization together to co-create solutions like a singular vision that can act as the north star.

When asked in a survey what topics C-suite level executives discuss most frequently as an organization, 57% of survey respondents put developing and creating new products at the top. Many blame failure to innovate effectively on organizational inertia—focusing on protecting the current state while avoiding or delaying responses to marketplace disruptions. An organization that is not siloed and is structure to bring people together so they can collaborate, be inclusive, and co-create is in a better position to overcome the organization inertia referenced earlier.

The third guiding principle speaks to the age-old divide between business and technology—a technology centric framework alone will never be an answer. Technology is a means to an end, not an end in itself. The starting point for any technological solution needs to be the customer or end-user.

Sometimes in a race to become a technology-driven organization and making use of new advances in technology to gain a competitive advantage, it's easy to forget the greater purpose, that is, to better serve customers while evolving with the marketplace. And this is where design thinking comes into play.

Design thinking is a human-centered approach to innovation that fundamentally rests on two essential principles: Empathy and objectivity.

Empathy

In order to innovate, you must understand the people you're trying to serve, the environments and contexts in which they operate, and the interactions they have with the world. Innovation all begins with observation: Through empathy, you can identify the problems people experience today as they try to get "jobs" done in their lives or work. It's important to document your observations carefully: Notes, photos, videos, interviews, customer quotes. These should be compiled in a diary that will help you understand what the customer "Does," "Thinks," and "Feels." In turn, these observations will give you insights that can help you identify and prioritize opportunities to innovate.

Objectivity

In this phase, you also want to keep an open mind and avoid being prisoner of assumptions that might blur your thinking. The challenge for leaders is to act as anthropologists who immerse themselves in the user context to see the world from their perspective—and understand reality as it is, not as they wish it to be.

At the end of the day, design thinking is focused on having an end-user perspective for value creation and people orientation at the center of their approach, and it's pivotal to prevent us from ending up with technology centric frameworks that are far from our customer needs. Instead, we need purpose-centric frameworks that are cutting edge and are focused on serving a customer need in the marketplace.

While we are on the subject of design thinking, I would like to touch on another competency that is fundamental in establishing a place where solutions get co-created: Inclusive and transparent decision making. Whether an org is transparent on how decisions get made or transparent on the actual decisions are two very different concepts and realities. Even socializing the 'how' of decision making can be powerful and engaging. And what we hear often is that keeping everyone up on the actual decision making in a growth atmosphere is almost impossible to do.

It's not impossible.

Being transparent about your decision-making process builds *contractual trust*. Communication trust calls for honest and frequent communication. When people advocate for transparency, what they typically care about is not knowing every detail. We quickly realize that's impossible. But rather they care about understanding how and why the decisions are made, and how and why they can participate in iterating on the actual solution.

Another key point to remember is that those impacted by decisions are often the best to determine whether those decisions were effective. Did we end up getting the desired effect? And within that, maybe retrospectively thinking how the decision-making process could be improved. This is the "why" that people most crave. The "why" is scalable because it carries the original intent of the decision and not just the tactic. And then we enter into a partnership to continuously improve on those decisions.

It's the combination—trust and collaboration—that turns the typical hierarchal decision-making model into a partnership. And that lets information sharing scale.

To accelerate a culture where solutions are co-created keeping consumers in mind, payers will need to revisit what is the value they are creating for their consumers, how they deliver this value, and by bringing people who create this value together and closer. Catalysts in this framework infusing collaboration are strategic alignment, consistent processes, and transparent decision-making.

* * *

Innovation, Simplicity, and Mindfulness

Richard Gannotta

Innovation is frequently characterized as an end product of an ever-evolving set of complex interactions leading to a new, more efficient, effective, and value-added product or outcome.

Although this notion holds true in many cases, deeper analysis reveals that effective system design usually is at its core "simple."

In biological systems, the notion of "emergence" where individual (simple) components of a large system work together to establish complex supporting behaviors (think an ant colony or flock of birds) is a well-established example. The "survival" and advancement of the "whole" is predicated on this collective behavior.

In some ways the same can be said for the pursuit of applications and new approaches not yet fully developed that are intended to solve or remedy some problem or challenge.

Starting at the base level, the most deconstructed aspects of any approach can create a path forward where efficient and efficacious development can occur without the friction and drag that may be associated with a complicated calculus.

A key example from healthcare can be found in instances where human factors have been identified as a significant issue in safety events. The interaction between an individual and a set of complex variables (technology, biology, psychology) can lead to an overwhelming and sometimes cascading set of events that can lead to a significant adverse outcome.

Is there a relatively simple yet highly innovative solution that can focus an individual's attention in those situations where high-reliability performance is required and reveal an emerging set of options and alternatives that can lead to better outcomes?

One such solution is the application of a "mindful" approach, which can serve to focus and frame a significant task at hand while objectively recognizing one's past experience or bias in a way that does not unduly influence the performance of the task or interaction but instead enhances and highlights areas where hidden patterns or variables can (not unlike the biological process previously described) emerge and be identified and untoward situations avoided.

We know this practice has "high utility" in complex areas such as medication administration, surgery, and other healthcare areas, but also air traffic control, nuclear power, and aviation. All examples where small negative variances in performance can have catastrophic impacts.

In the realm of technology, the same holds true, and the need for a mindful approach when inputting data, performing successive or rapid keystrokes, and analyzing information given the potential impact faulty conclusions can have on system outcomes is essential.

A quick approach to applied mindfulness as it relates to interacting with technology starts with the following:

1. Mindfully "check in," anchoring yourself in the present moment through breathing awareness and subject focus then engaging the task at hand
2. Prior to entering data

3. When interpreting information displayed in complex formats
4. If display fatigue may be possible
5. In any instance where interfaces between technology and biological processes yield actionable data

So how do we adopt a mindful approach to unlock the simple side of innovation allowing for a more efficient and value-added approach to "emerge"?

Organizations should consider adopting a culture of high reliability powered by a formal mindfulness program.

The end product could not only produce a safer more reliable environment but also empower individuals to identify and deploy corrective measures in areas of vulnerability and experience-enhanced teamwork and to be more present in the moment, allowing for the emergence of new innovative approaches to solving complex processes.

* * *

Innovation in Healthcare Information Technology: Stressing Simplicity

Adam Buckley

Innovation. Revolution! Upheaval! Radical, industry rattling change! The word itself denotes such an inexorable alteration of the landscape, it is hard not to get swept up in the possibilities of creating that next great disruptor. In 2017, we undertook a complete redesign of IT services at our parent academic medical center. As part of that work, I envisioned a research and development team that would be wholly within the "innovation space" and create nothing but new and exciting ways to deliver care. The fact that I planned to monetize their work via patents and licensing also created a great deal of excitement within the halls of finance. So with my vision in front of me, I recruited and put together a crackerjack team. My next move was to talk with people I knew who had lived in the innovation space. I have some friends who have made crazy money in healthcare IT (HIT) and had real-world experience in exactly what we planned on doing. So I put together my pitch deck and my elevator speech about all the work done to date, ready to hear their secret to success. Imagine my surprise when it came down to just four words. Solve a real problem. That's it, I thought?

They had made untold millions on that? My next thought was, why did I go to medical school if it is that simple to improve the lives of patients? With my renewed calling to focus on solving real problems for our patients and providers, my team and I refocused our efforts.

The team I created had individually created real value for the organization over a number of years by producing niche applications that various parts of the organization had asked for. Most of these existed in gaps between what vendors offered in the commercial space and the need providers and operations had doing their day-to-day work. Many of these were also solutions that could be homemade as opposed to spending vast sums to acquire. One of the more successful creations was a mobile application that gives patients' families the chance to see where their loved ones are in the process of having a procedure performed in our largest facility. Is the patient in holding? In the operating room? In recovery? This was necessary since the facility had three IT systems to perform the work of supporting admitting, the operating room, and the postoperative and in-patient stay. Patients have grown to love this application. It solved a simple communication issue for the staff and yet gave patient's families vital information. Simple. The team also developed an emergency desktop notifier that popped up whenever there was an issue that required mass notification. When this was developed over ten years ago, there was nothing on the market to help. It was built not at the request of IT, but from facilities. They wanted to know how to communicate broadly to all employees about widespread issues. The organization uses it today for a variety of simulations (active shooter, baby abductions, etc.) as well as facilities issues and IT issues. It has become one of the many ways that we contact staff in an emergency. Simple.

With the past successes as a starting point, we decided to move ahead. That began with securing a lawyer with expertise in intellectual property. With their input, we developed a process whereby the team could document while they developed in a way that lent itself to a patent application should their work bear fruit. We also began a process to perform suitable due diligence to receive provisional patent protection as the work continued through alpha and beta development. This is a vital step. Receiving full patent protection is a long, painful process that requires a significant commitment of time, money, and resources. It also requires very specific documentation that calls out the novel and new way that you are approaching an existing problem. Provisional protection at least offers an opportunity to develop and not worry about someone else stealing the idea. One can then determine if a full patent application is worth it after the application is fully developed and the concept behind it fully realized. We have actually received such protection on a few applications that we believe may eventually be licensable.

Our next step was to integrate our existing internship program with the new development team. We require our intern teams to produce an application by the end of their internship. One team produced an application that calls up maps of our various facilities and helps facilitate patients finding their way through the myriad 100-year-old buildings that we have on our campus. Historically, the organization produces an oversized novelty map that is impossible to read at great expense annually. The map project, while not patentable, produced an application that has the potential to save money for the group. Again, simple.

The most recent step has been to integrate the new development team with our newly formed clinical innovation lab. The innovation lab is charged with looking for opportunities to map out new methods of delivering care in a "value" -based reimbursement model. This lab is focused currently on filling the gaps in care that our health network has with referring providers. We are in a rural region, and many referrals come from out of state or great distance, and there are no existing means to exchange information easily. Many health systems such as ours have experienced the disappointment that despite leveraging eHealth Exchange and other methods of exchanging clinical information, there is a still a significant gap in what is provided via the exchanges and what providers and patients need. This effort is currently ramping up as a pilot, and we have a provisional patent, so the excitement over the work is building. However, it is also trying to solve a simple problem: Helping providers and patients communicate about their care.

We have much to learn as we continue in the world of development and innovation in healthcare. There have been failures as well as successes. That being said, my belief is if we stick to the simple premise of finding solutions to real problems we will succeed in the long run.

* * *

Stress Simplicity in Innovation

Cynthia Davis

As a nurse, technologist, leader, and caregiver, I am most excited about the ongoing transformation of the delivery of health. Throughout my career, my focus has been on identifying opportunities to improve the quality and outcomes of patient care delivery with a strong foundation of effective cost management, process, and systems of care. We are at the first ten-mile marker of a journey of 10,000 steps that will take curiosity, collaboration, and innovation to support provider/clinician and consumer transformation.

Here is an overview on how my organization stresses simplicity in innovation in everyday practices of care.

Methodist LeBonheur Healthcare (MLH) is a six-hospital integrated delivery health system serving the Mid-South region. We are celebrating our hundred-year anniversary serving our community. We do $1.92 billion in revenue and see 405,000 ED visits, nearly 64,000 hospital admissions, and 1.4 million ambulatory and home care visits per year. Our promise is to improve every life we touch through our organization's values. What that means for us is that we are a learning organization and embrace new ways to get better results. This includes the following guiding behaviors:

- I am personally willing to change.
- I am curious and openly seek new approaches, processes, technologies, and practices to improve outcomes.
- I collaborate with patients, families, and my team to implement new ways of improving the healthcare experience.

Like many other healthcare delivery organizations, MLH invested early in health IT and enhanced those capabilities in part through funding from HI-TECH. Those EHRs now contain quantitative data (e.g., laboratory values), qualitative data (e.g., text-based documents and demographics), and transactional data (e.g., a record of medication delivery). The big benefit with the EMR is the aggregation of patient data over the long term.

With the foundation in place, this phase of care innovations enabled by technology now allows for a transformation of care by delivering information directly to patients and empowering them to play a more active role in their care. We're now exchanging data—not just for meaningful use but to improve patient health. We're connecting with Methodist partners and physicians in the community (including providers who serve both underinsured and self-pay patients) through mobile applications and text messaging programs that provide disease-specific education and reminders.

We also have cutting-edge initiatives underway, including joint projects with Cerner, Big Data work with the University of Tennessee, and precision medicine and population health programs—all of which are aimed at serving a unique, underserved patient population.

And in fact, it was that culture of innovation that led to the development of a simple life-saving alert system based on Cerner tools. With 90% of doctors using iPhones, iPads, or Apple Watches, the IT shop created an algorithm that sits on top of its EHR platform to continuously monitor changes and alert clinicians accordingly.

This algorithm has already saved thousands of lives by shrinking the time it takes to notify a physician when a patient is diagnosed with severe sepsis from six hours to fewer than five minutes. It also alerts physicians when a patient is at risk of fall or readmission; users can hover over the alert for clinical decision intelligence that explains the steps that can be taken to avoid readmission.

Our EHR vendor, in turn, has added that algorithm for other hospitals to use. It's been amazing to see an innovation like this take hold.

Here are some key takeaways to enable patient-centered roadmaps of innovation:

1. **Maintain organizational leadership and develop blue chip-focused support structures**. At present, many organizations are trying to manage enormous numbers of requests for IT changes. If these are not actively addressed, value is not likely to be achieved, with the consequence that "sharp-end" providers may become discouraged. The systems that vendors offer tend to be "bare-bones," and the implicit assumption being made is that organizations will use the system tools offered to make care improvements. But achieving this requires organizations to continuously develop local human resource and governance structures and focus on key blue chips and priorities for patients and families.
2. **Look for low hanging fruit and focus on small, simple improvements**. Innovation in so many ways can be the simplest of things. We just need to be doing it more successfully and efficient than anyone else, working the most effectively with what we have. The simple act of chewing an aspirin has shown to improve clinical outcomes from an acute heart attack.
3. **Improve the basic care and business process first**. Before taking on new initiatives, focus on improving today's business or clinical problems. Success breeds success. Agreement in relation to goals is important to ensure that optimization efforts of stakeholders are aligned. Look for small, simple improvements, and continue to improve.

4. **Promote transparency as a cultural norm**. If we don't promote transparency of results, outside regulators or payers will do so, and not very gently. We have the data in our EMRs. Those leaders who promote this will find that the benefits outweigh the risk. The more the data are used, the better it will get.

5. **Create a top-down vision and stimulate bottom-up innovation**. Actively managing the process of change is essential because all organizations have difficulty in navigating major organizational change. Effective organizational transformations require long periods of time and constant effort.

6. **Set a specific benefits-driven approach**. Start with a benefits-driven method. Our approach begins with the identification of a specific aim (e.g., 5% reduction in hospital readmissions within 12 months), followed by an assessment of current and future states. After these important preliminary steps, relevant data items are identified and specified, which allows monitoring of progress toward this goal.

We're seeing small and simple innovation take hold and produce results. Advancing health outcomes and ensuring that healthcare fits seamlessly with an individual's lifestyle is our key focus.

* * *

Stress Simplicity and Collaboration in Order to Accelerate Change

Kevin Dawson

Healthcare is a conservative industry and a late follower of technological innovation. In 2023 healthcare companies still rely on yesterday's technologies (e.g., faxing, printing, and scanning). In healthcare, some stakeholders perceive a difference between two types of responsibilities, one that is performed without technology and another one that relies on the use of technology, for example, interaction with electronic health records (EHRs). Statements such as "I'm not computer literate" and "I was not trained to do data entry" are still voiced by healthcare professionals resisting technological change. In contrast to healthcare, in most other industries, these attitudes against technology adoption and change have not been seen since the 1990s.

Conservative stakeholder attitudes against change is not the only cause hampering technology adoption and innovation in healthcare. The high complexity of the U.S. healthcare system with numerous providers and payers led to the evolution of overly complex EHR systems. While the EHR market for large acute care hospitals consolidated to a couple major players, it is not unusual to see as many as 150+ clinical and non-clinical software applications in a healthcare organization. Many of these systems may look modern on the surface but, in fact, rely on old technologies in their cores. Even healthcare organizations using the same main EHR may have strikingly different EHR implementations and portfolios of best of breed applications.

Innovation often targets improvement of efficiencies. However, healthcare is typically not incentivized to pursue efficiency. In many contexts, the Certificate of Need (CON) requirement creates a non-competitive landscape. The CON requirement keeps new entrants out of the inpatient healthcare market even if they may have novel innovative ideas on how to perform healthcare services more efficiently. Moreover, healthcare is a risk-averse industry.

Anything new is perceived as risky, which leads to the perpetuation of "tested" but inefficient solutions, for example, faxing. The complex reimbursement system also favors old ways of delivering care. This limitation can be seen when delivering telemedicine consultation. Even if patients and providers both prefer telemedicine consultations over in-person visits, providers must ensure they are reimbursed by payers before implementing a new telemedicine service.

The high complexity of the U.S. healthcare system and the lack of incentives for technological and business process innovation and adoption are key reasons for why the U.S. healthcare system is the most expensive of all. Despite the extraordinary costs, the U.S. health outcomes are only mediocre at best. While politicians on both sides of the political spectrum work toward improvements to address this economic challenge, the gap between healthcare cost and outcome doesn't seem to be closing anytime soon. While technology is not a silver bullet solution to address all these challenges, I argue that technological solutions and business process innovation may contribute to the solution.

I also argue that technological innovation and adoption of change can be implemented in healthcare organizations only if close collaboration exists between clinical, business, and technology leaders. This collaboration can be done effectively only if technology solutions are deployed in a streamlined, easy to use, and efficient way. Reducing complexity is key to better and faster adoption of innovative solutions. The best solutions allow users to perform the common tasks easily while making the more complex tasks also possible.

In the spring of 2020, the U.S. healthcare system responded to the COVID-19 pandemic. Patient volumes increased exponentially without a clear foresight of what the maximum peak of the patient surge would mean and with no definitive treatment or prevention of the disease on the horizon. Resource constraints surfaced right away for nursing staff, hospital beds, intensive care unit capacity, emergency department capacity, oxygen, medications, ventilators, personal protective equipment, morgue capacity, computer equipment, and many more. At the same time, various government and private organizations made resources available to healthcare organizations. One of the memorable signs of the time were the free meals delivered by local restaurants to healthcare workers.

This case study describes the deployment of an inpatient telemedicine solution in an urban, non-profit, academic medical center during the COVID-19 pandemic. The organization historically serves a patient population that is uninsured or insured by public payers such as Medicare or Medicaid. The hospital is owned by a private university; serves as the teaching hospital for the medical, nursing, and pharmacy schools; and hosts multiple medical residency programs. Most of the hospital patients are admitted through the emergency department and present with high acuity illness, often with many comorbidities. The combination of a more severely sick patient population and limited reimbursement from public payers has been financially challenging and led to chronic resource constraints and high senior leadership turnover. Despite these challenges, many Union-represented employees have been highly loyal to the organization and have worked there for decades.

In response to the COVID-19 pandemic, the hospital initiated an accelerated reconstruction program to open more departments and convert existing patient rooms to multi-occupancy rooms in a couple of months. Furthermore, the hospital opened an emergency department expansion in a tent. Plans were developed for opening additional medical/surgical units in tents and in a nearby convention center. The expansion work was performed under a very tight deadline, in anticipation of an imminent, major COVID-19 surge. The new departments required fast deployment of a nurse call system, wired and wireless networks, medication and supply dispensing cabinets, and configuration of the new space in the EHR. At the same time, as many hospitals worked toward the same goal, multiple supply chain issues became apparent right away.

Parallel with the accelerated expansion, the hospital applied for a federal grant to deploy an inpatient telemedicine solution and medical tablets for patient entertainment. The justification for the inpatient telemedicine solution was to provide care remotely to patients in the hospitals by physicians from outside of the hospital. While adequate space and physical resources were developed for the COVID-19 surge, staffing of the new departments remained a major challenge. Fortunately, many recently retired physicians considered returning to work from retirement. However, due to their retiree age, they were more vulnerable to develop severe disease in case they contracted the COVID-19 virus. At this time, no vaccine or specific treatment was available. Being able to see their patients remotely was attractive for these physicians. Moreover, during the pandemic, physician licensing rules were relaxed.

Physicians with an active license in any of the 50 states were able to practice medicine in the hospital. Therefore, an inpatient telemedicine system provided the hospital with more physician recruitment options.

The tablets served the purpose to connect patients with their families, particularly in the isolation rooms. During the pandemic, initially, the hospital didn't allow visitors, with very few exceptions. Consequently, patients felt isolated, and patient satisfaction declined. The tablets allowed patients to communicate with their loved ones while in the hospital. The medical grade tablets were shipped in plastic cases that were easy to disinfect and clean. Moreover, the tablets had RFID and bar code readers for another purpose to use them as EHR endpoints. Cleaning a tablet is much easier than cleaning a workstation on wheels (WOW). Therefore, many nurses preferred entering an isolation room with a tablet rather than a WOW. The barcode readers and RFID readers can be used for reading patient wrist bands, employee IDs, and medication barcodes.

In addition to purchasing the telemedicine equipment mounted on mobile carts and the medical grade tablets, the hospital also invested in its wireless network to support the new wireless equipment. The telemedicine carts had high-resolution cameras and required high quality wireless connections. After the receipt of the federal grant, the hospital had only couple months to purchase and deploy the new equipment. While the deadline was later extended, the hospital was successful in deploying the equipment by the original deadline. This was largely due to a long-term partnership with a local IT infrastructure firm that was able to scale up capacity and provide resources by a very tight deadline. Furthermore, in order to deploy the tablets by the deadline, they were first shipped without the RFID readers and were deployed right away. Later, batches of tablets were terminally cleaned and sent back to the vendor for the installation of the RFID readers. This process allowed the hospital to start using the tablets without having to wait on the rate limiting step, that is, the installation of the RFID readers.

Deployment of the new technical capability was just part of establishing a new program. Creating a new service line required collaboration among multiple departments, including physicians, nurses, and IT. Parallel with the technical work of deploying the wireless network improvement, telemedicine equipment, and tablets, the new services were also defined. While most physicians were excited about the telemedicine program, part of hospital leadership raised multiple concerns that had to be addressed in advance of the service rollout. There were concerns raised about how telemedicine visits would be documented in the EHR, how patients would be consented to the visits, whether the hospital would be reimbursed by payers for the visits, whether some clinical specialties may opt for telemedicine consultation just out of convenience and may risk quality of service, how clinical residents would be supervised, how third parties (e.g., family members, interpreters) may be included in the telemedicine visits, potential patient privacy risk in semi-private rooms, what training is provided and by whom for physicians and nurses operating the equipment, and whether consultation with the Unions is required before tasking nurses with operating and disinfecting the new hardware in the hospital.

Similarly, nurses were given the new responsibilities to issue, collect, and disinfect the new tablets and sometimes had to answer technical questions raised by patients. In case a tablet didn't connect to the hospital's guest network as expected, nurses needed to work with the hospital's IT department to resolve the issue. While policies were published for all these tasks, managing the new equipment was considered a lower priority than direct patient care, and therefore tablets were often not issued to patients. The hospital didn't want nurses to perform non-clinical tasks due to a nursing staff shortage and the very high cost of temporary nurses. After the hospital recognized that nurses did not issue the tablets, many of the tablets were repurposed for other uses. Most importantly, they were used as EHR endpoints, particularly, in isolation rooms.

The extent of utilization of the inpatient telemedicine service was less than anticipated. The most important reason was that the COVID-19 surge was fortunately much less severe than anticipated. By the end of 2020, when the telemedicine program was deployed, the COVID-19 patient volume was declining. Moreover, hospital leadership restricted the use of telemedicine service to specific departments and by specific clinical specialties in order to prevent its use out of convenience instead of providing the service in person. Later, additional use cases were established, particularly in the emergency department.

In order to promote the use of the new capability, the services were implemented in a straightforward, easy-to-use fashion. Clinical documentation was performed outside of the telemedicine service, using the same tools physicians were already familiar with. Patient consenting was executed on paper to align with the process utilized throughout the hospital.

Although this seems to be a step backwards, it simplified the rollout of the service and made it more palatable for nursing staff and the HIM department. Tablets were also configured in a way to support ease of use. These simplifications were necessary for improving the utilization of the new capabilities. While the new technologies could have performed much more, their functionalities were reduced in order to support their acceptance.

This case study demonstrates the need for securing support from all stakeholders for a new program in advance. While all members of hospital leadership voiced their support during the application for a telemedicine grant at the time when the COVID-19 pandemic reached the United States, some of them were surprised and raised their concerns only when the program was eventually funded. While all concerns were addressed, the program was still limited by some external and internal reasons. Fortunately, the extreme COVID-19 patient surge, what the hospital prepared for, did not materialize. Furthermore, healthcare's risk-averse and late technology adoption culture prevented a full-fetched roll-out of a new telemedicine program. However, by simplifying and sometimes redefining the services, the hospital was able to deploy them. During the period after deployment, the hospital saw improvement of the patient satisfaction scores. The new telemedicine capability and patients' access to tablets contributed to the improvement of patient satisfaction at this time.

* * *

Simplicity in Action—Digital Pathway for Breast Unit

Carla Samà

What Is Digital?

What do health and space have in common? More than we think. Even the taxpayers' money, as in the healthcare sector, is spent on space travel, so no one wants to, nor can afford to, spend it irresponsibly. The duty is: Maximum performance, minimum risk. *It all has to work.* In recent years, the approach has definitely changed: The space industry has been joined by the digital economy (think Elon Musk's SpaceX). Now, to carry 1 kg of material into space costs 2,500€; before SpaceX it cost 50,000€. This is extraordinary.

The logic of the digital economy is the same that leads the market to release a new cell phone. When we buy a cell phone, we already know for sure that after five to six months, the same cell phone will come out in a better update and with a more powerful version, but this does not prevent Apple or Samsung from introducing a cell phone into the market where it will then be improved. This aspect has also changed the way we go into space. Today, in fact, we talk about "the new space."

This is digital! We always discuss with our colleagues, if it is working in space, it will also work in healthcare. It's an engineering approach, a completely different risk. You have to stay side-by-side with doctors, nurses and/or patients, grab their attention, see and study their reaction, then improve and adapt the product (possibly change it) to have more and more positive impact.

Before telling you our experience, just let me share two suggestions and a proposal approach.

First suggestion. Start and adopt. Integration at all costs for its own sake serves no purpose. GE spent eight years, 3,000 programmers, and more than $7 billion in an effort to develop its digital transformation software platform into which to integrate all the information available in the company (a space enterprise) GE Digital—an effort that ultimately contributed substantially to the company's failure and the replacement of its iconic CEO.[1] Today it is important to start. And then adapt iteratively. Then eventually change. Time has a cost to the health system. We have to do it quickly. We have to take risks. Systems that cannot evolve by integrating what is needed to be changed will die!

Second suggestion. Done is better than perfect. Non-perfection is a value. The rate at which innovation runs to approach digital implementations as a sequence of defined steps and waiting for the perfect finished product carries the risk that the product is already born "old." In order to digitalize the patient pathway of breast cancer, we have a platform (a web-based service and an app); maybe not the perfect one, but we need to address its functionality based on the priorities of physicians, nurses, and patients. And give an avenue through which they can share their needs.

Let's come to the idea of approach, based on what we applied to deal with the complexities. We need to be able to distinguish these three situations.

The first one: Does it serve the clinician that we work with and attempt to simplify? The example is the inclusion of the team as the provider of the service in the compilation of the individual care plan. Theoretically, the clinician who will perform the service is expected to be listed, but often we don't have that information. We don't get stuck and have them enter the team. You don't know what the team is? Put something in, and we'll prompt you that this information may not be usable.

The second: If the patient requests it, we go deeper and analyze the possible spaces of applicability. Think about televisit. In a study published in *Surgical Oncology*, it revealed that 85% of breast cancer patients liked the service delivered by telemedicine, 66% would do it again, and 79% would recommend it to a friend or family member. Another study in *JCO Oncology Practice* surveyed Breast Unit physicians, and 92% of respondents agreed that telemedicine can improve the patient's care pathway, and again 92% believed it can be valuable in the early stages of follow-up.

Third: Do not spend time to make perfect features of the system that neither the physician nor the patient needs. Think about many procedural aspects. Study them, arrive at a solution quickly, and then stop talking about them. Perfection is a pointless pursuit. Do what it takes to drive the evolution of the platform by focusing on what users and patients need.

An Approach for Breast Unit Care Pathway Digitalization

The Breast Units digital transformation project we want to talk about aims to virtualize the clinical network by uniting roles, places, and knowledge through which the patient journey unfolds. The goal is to enable a new model of care through the collaboration of professionals around the history of breast cancer patients, from screening to follow-up. Our client, a big Italian region, has equipped itself with a platform choosing to start with Breast Units.

The failure of digital transformation initiatives is often linked to the adoption of an approach that sees transformation as merely a technological aspect. Instead, technology, however enabling, has only a small impact on the success of initiatives, linked primarily to people and processes. If we were to express it in percentages, digital transformation is 80% people, 15% processes, and 5% technology.[2]

Based on our experience, we have gathered some recommendations that help the digitalization journey.

Listen to dissent: Epic, a leading American software company in health care, in 2002, saw the failure of a project to insert electronical medical records at Cedars Sinai Hospital in Los Angeles despite a solid technological foundation. They tried again a few years later, listening to every single voice until all the questions had been answered, and the same project achieved the expected goals. In approaching the Breast Units, we observed that during the kick-off meeting, after a brief alignment of why we were there, listening to the critical issues encountered in the daily work (e.g., lack of staff, inadequate technological equipment, insufficient time to devote to the patient due to excessive bureaucracy) was a step that facilitated building empathy and engagement with the users.

Create a sense of urgency: Establishing a sense of urgency or need for change is crucial to gaining the necessary collaboration. In order to do this, the sponsorship and commissioning of the management and the involvement of all Breast Unit constituents were crucial. Breast Units are engaged with dedicated kick-off meetings to motivate importance and value of the initiative. Why now? Because we need a digital solution that can improve quality of care and the daily work of healthcare personnel. Why the Breast Units? Because cancer is the emperor of all diseases, and breast cancer is one of the cancers that can most serve as a pathfinder for other types of disease because of the emotional impact on the patient and the history of the disease.

Ensure to close the care gaps: "Connect the dots" in the continuum of care. To enable physician adoption, the digital platform has to move frictionless around other information systems that already manage data on the patient journey. When we think about breast cancer, our platform must allow to view, query, and receive information and alert taking data from at least the registry of the patient, the electronic medical record, radiology system, anatomy pathology system, laboratory system, and screening database.

Make technology transcendent: We call it transcendent products. It's essentially universal design that has been dialed up to 11 on a 10-point scale, with accessibility attributes so useful that they turn out to be highly desirable—even aspirational—for all people.[3] Achieving ease not by patronizing one's customers but by exciting and delighting them is a strong lever to digitalize processes.

Act on short-term wins: Carrying out a transformation without having attention to short-term daily wins is extremely risky. Acting on small, successive improvements fuels confidence and the drive for change. In engaging Breast Units, one of the recurring concepts is that the path is neither easy nor linear and that we need to act by successive improvements, ensured by the flexibility of the technological solution to other domains. Then again, if we went back in time today to the first version of Amazon, we would be amazed at how basic it was. In the case of Breast Units, we believe that an example of short-term wins is to allow the multidisciplinary team to perform on the platform by digitizing pre- and post-surgery multidisciplinary records.

Encourage adoption: We noticed that healthcare personnel when joined at training times in the use of the platform, partly because of the platform's user interface, showed themselves to be interested, engaged, and good users. However, as the days went by, we noticed a curve of unlearning and thus disaffection. To encourage adoption, we pushed physicians and nurses by sending them digital contents and notifications that could remind them of the platform's features and keep them engaged. We acted following the Kotler's 5A logic (Aware, Appeal, Ask, Act, Advocate).[4]

Push informed decisions for better outcomes: Digital transformation has to enable a joint and integrated rethinking of the goals of upgrading the patient experience in-together with those of clinical governance of the complex mechanisms of the NHS. We believe that a bidirectional interconnection must exist: The patient experience is facilitated by digital tools, practitioners (who are also Key Users) are facilitated in pursuing their specific purposes, and at the same time, data and information arise on a management level that integrates various sources and reads information in a unified way. From this principle, the level of government derives general guidance to guide the health service as a whole or in its individual articulations and simultaneously returns to the patient and providers digital tools and data through which more informed decisions can be made to protect outcomes and system sustainability (e.g., the risk of a patient abandoning a therapy, or the risk of a patient being readmitted to the emergency department).

Telemedicine and Instruction for Use in Breast Cancer

The COVID-19 pandemic has spawned an unexpected boom in telemedicine: The Cleveland Clinic, considered a leader in telemedicine, went from 2% to 80% of outpatient visits delivered by telemedicine during the pandemic, to stabilize at around 13% these days.[5] Stories like these have been replicated in many settings, including in Italy. The Italian recovery and resilience plan dedicates 1€ billion of investment in this field to foster the adoption of vertical telemedicine solutions. Guidelines[6] have also been issued that define functional and technological requirements for different forms of telemedicine.

Those who experienced breast cancer from professional experience or unfortunately from user experience know that this form of cancer is characterized by "high tactile" involvement of the patient. Adding to this critical issue that constitutes a clinical barrier to the use of telemedicine is a poor cultural predisposition of health care providers. The idea of starting with breast cancer, however, lies in the fact that cancer is the emperor of all diseases[7] (it covers all stages of the value chain and the complexities of medicine: Prevention and screening, medical treatment, surgery, rehabilitation, follow-up, chronicity, vaccines), and breast cancer is the most frequently diagnosed between cancers.[8]

How can the adoption of telemedicine be fostered starting with oncology and Breast Units? To this question we answered, "Let's tackle it head-on!" How? (1) By identifying the areas of applicability of telemedicine and formalizing them into guidelines by helping with what has been done elsewhere; (2) by helping physicians and health care providers acquire the semantics (e.g., wording, new behaviors during the visit) necessary for telehealth to be successful with the patient; (3) by verifying that the physical-technological conditions are in place to deliver it; and (4) by raising awareness about the possibility of having telehealth available to patients (first and foremost) and health care providers.

Telemedicine Applicability Spaces, Shared Protocols, Analysis of Best Practices

Although guidelines governing the use of telemedicine have been issued at the national level, there are no widespread operational protocols for the application of telemedicine to specific pathways.

To address this issue, we think there is a need for the shared drafting of an operational manual through a working table composed of professionals from inside and outside the Apulian oncology network, with both technical and clinical expertise that is supported by system stakeholders. Such a document would not only make it easier for physicians to choose to apply a service remotely but would also address the need to formalize the clinical guarantees and operational requirements required by national guidelines. In order not to reinvent the wheel, there are national and international experiences to build on.

Helping Physicians and Health Care Providers of Today and Tomorrow in Acquiring the Semantics Needed to Deliver Telemedicine

By 2021, more than 73% of physicians and 57% of nurses used messaging apps to deliver healthcare advice and services.[9] From the latest available research, only one in three health professionals use dedicated or certified communication platforms, although interest is especially high among physicians (74% of specialists and 72% of general practitioners): Handcrafted solutions that worked in the pandemic emergency but risk undermining the value of telemedicine. Assuming then that we can translate the logic of physical performance to digital is a mistake to be avoided.

To do so, it is necessary to support providers in learning the necessary new semantic techniques for managing a virtual visit, to include in providers' training courses dedicated to learning the techniques and modalities of new virtual forms of care (e.g., remote monitoring, tele-board, tele-consultant, tele-ICU, televisit), and to promote studies and practical advice for telemedicine providers.

Adapting Hardware Infrastructure, Software, and Space Layout to Telemedicine

The American Society of Clinical Oncology (ASCO)[10] identifies among the benefits of telemedicine are a reduction in time spent commuting, immediate access to care, and a reduction in caregiver and patient burden. In addition, the Digital Health Observatory[11] estimated that through increased use of telemedicine services, it would be possible to save about 48 million hours to date wasted in avoidable travel, which rises to 66 million hours when considering that 35% of patients are usually accompanied to the facility by a caregiver.

Despite this, survey results highlight that the real development of telemedicine is hampered by critical factors related to inadequate physical and technological infrastructure. In particular, a survey conducted by ALTEMS[12] reports that only 5% of participating healthcare providers consider their organization's infrastructure adequate. Many users, in fact, experience, on the one hand, the lack of space, dedicated workstations, and/or technological tools for conducting a televisit (e.g., one hospital did not have webcams or microphones) and, on the other hand, the technological backwardness of the tools already available (the PCs of many facilities still have Windows XP as their operating system, released in 2001).

Therefore, a critical element of paramount importance is to verify that operators have the technological and physical infrastructure to deliver tele-medicine (intra- and extra-hospital). To do this, it is necessary to conduct an assessment of the actual presence of the devices, and their status, required to perform telemedicine services, as well as the level of connectivity of the workstations (e.g., access to the intra-and extra-hospital network and Wi-fi connection). In addition, collaboration with facility managements to reorganize the layout of spaces would help to redevelop those that are unused, or misused, to create environments dedicated to the delivery of telemedicine services.

Increasing Patients' Awareness about Telemedicine Services

Many citizens have experienced telemedicine, often provided in a hand-crafted form, during the pandemic. With the emergency over, however, we are not seeing a massive phenomenon of demand for telemedicine, even though the pandemic has changed our behaviors, making them more akin to wanting to seek more and more digital experiences.

Having therefore verified the adequacy of the IT infrastructure, the technological requirements, and the way in which the use of telemedicine services is validated, it will be necessary to focus efforts on widespread communication with citizens so that they are not only aware of the services and benefits but also request, use and recommend them.

To do this, it is necessary to spread awareness of the existence of the possibility of telehealth within facilities and to exploit the communication channels closest to citizens to increase patients' awareness of available tele-medicine services ("I know the hospital offers telehealth") and also readiness ("I am ready to answer when the doctor calls").

Agile as a Way of Life—A Change Management Framework

Digital is not a project with a start and finish (traditional). It is a journey and should be planned and executed as such (agile). We used a framework in order to make the grounding of the platform stick in the breast units. The framework is one framework (not the only framework), but it allows us to have a guide that can be scaled up also for other pathologies and geographies.

In the "make it clear" phase, the goal is knowing the peculiarities of the single Breast Unit and its team members. The best antidote for not allowing any kind of transformation to be derailed is to listen to dissent. So, in approaching a new Breast Unit, we focus in understanding the way it works, collecting feedback from people about the platform, and explaining the initiatives.

In the "make it known" phase, the goal is making clear the importance and value of the initiative, sharing with operators the process improvements that will be generated and the associated value added. In this phase, it is important to create a sense of trust; we generally create a common chat in which operators can write problems and receive support in using the platform. This chat is also used to send "smart pills" about the platform functionalities to enhance the features and remind clinicians how to use them.

In the "make it real" phase, the goal is supporting practitioners in enrolling patients, coaching on "field use" of the platform, supporting healthcare operators in addressing issues that arise during implementation, training on any new features added, and detecting improvement opportunities and new features.

In the "make it happen" phase, the goal is listening to patients during the care pathway. This phase is characterized by initiating a communication campaign with enrolled patients, collecting patients' impressions of using the app as a part of the platform, and launching a patient support program. As an example, we allowed the multi-disciplinary team to be carried out on the platform by digitizing pre- and post-surgery multi-disciplinary records. This allows the team to track and consult the patient's biographical and clinical information digitally in one place instead of doing it on volatile paper.

In the "make it stick" phase, the goal is making innovation lasting. In particular, the main activities regard measuring the impact generated on the organizational model, launching a user-satisfaction survey, sharing process and technology improvement elements, and supporting the use of new platform features. As an example, we developed a dashboard to share weekly with the stakeholders about the numbers of the new patient enrollment and the usage of the platform.

The framework can be graphed as a roller-coaster because it can require iteration as digital transformation is anything but a linear sequence, nor does it have a real end. As a demonstration, think of platforms such as Amazon: If you search for Amazon on the website The *Wayback Machine* (https:// archive.org/web/), you can find the first edition of the Amazon website, and you will be amazed at how different it was from today's version (see Figure 6.1).

Figure 6.1 **The agile approach.**

Notes

1 Digital Transformation—Survive and Thrive in an era of mass extinction—Thomas M. Siebel.
2 Healthcare digital transformation: how consumerism, technology and pandemic are accelerating the future, Edward W. Marx, Paddy Padmanabhan.
3 The longevity economy: unlocking the world's fastest-growing, most misunderstood market—Joseph F. Coughlin.
4 Marketing 5.0—Philip kotler; Kartajaya hermawan; Setiawan iwan.
5 Digital Voices—E. Marx.
6 Telemedicina, Linee d'indirizzo nazionali (20/02/2014), Indicazioni nazionali per l'erogazione di prestazioni di telemedicina (28/10/2020); Linee di Indirizzo per i Servizi di Telemedicina (Allegato B al Sub-intervento di investimento 1.2.3.2 della Missione 6 del PNRR) (30/09/2022).
7 The Emperor of All Maladies: A Biography of Cancer, Siddhartha Mukherjee, 2010.
8 Italian Cancer Registry Association—Cancer numbers 2020 link.
9 Research School of Management's Digital Health Observatory (Polytechnic of Milan).
10 ASCO Interim Position Statement, Telemedicine in Cancer Care, July 2022 (link ASCO).
11 Osservatori.net digital innovation, Sanità Digitale oltre l'emergenza: più connessi per ripartire, Osservatori.net.
12 Survey Alta Scuola di Economia e Management dei Sistemi Sanitari dell'Università Cattolica del Sacro Cuore (Altems) weblink.

Chapter 7

Recognize and Reward

Recognize or reward the efforts of stakeholders to innovate even at the smallest levels.

To maximize innovation potential, we must not forget the power of motivation in human behavior. People will largely do what they are primarily rewarded and recognized for. For innovation to thrive, consider launching multiple reward and recognition programs to reinforce culture, enhance engagement and encourage collaboration. Programs should reward not only those who generate ideas but also all the support teams enabling the success. That which is rewarded and recognized is repeated. Innovation will multiply commensurate with affirmation given.

Sustaining a Culture of Innovation

Edward W. Marx

We tend to repeat those things we are rewarded for. I have served with several progressive and innovative organizations. The key from a leadership perspective was sustaining our culture of innovation. There are plenty of forces at work that want to push back the boundaries and return to the status quo. If you are not purposeful in holding ground that you have captured, you will lose it to cultural inertia. It took us a while, but in each case, we realized that it was imperative to reinforce the new culture through reward and recognition. In fact, reward and recognition became the catalyst for sustaining our innovation culture at every organization.

DOI: 10.4324/9781003372608-7

We learned that to enhance the probability of success, reward and recognition were equally important to apply when we failed. Counterintuitive for sure, but one of my colleagues argued that celebrating failure was more powerful than celebrating success. So we did. Most individuals are at least modestly influenced by fear and failure. As a result, over time we become increasingly risk-averse. Aversion to risk is one of the primary enemies of innovation. Embracing failure casts out fear.

Perhaps my biggest innovation failure was investing resources in a product originally called Surface. The year was 2010, and there was nothing on the market like it. We built an interactive application that would enhance the clinician–patient experience. Less than one year later, Apple came out with the iPad, and the rest was history. Instead of licking our wounds, we decided to use Surface as a coffee table in our office. It was a daily reminder that sometimes we fail, but we are unashamed. As it turns out, that experience was a catalyst for us to develop an application on the iPad that would help diagnose Alzheimer's.

As a result of our rewarding and recognizing both success and failure, our innovation culture ran deep. When we had success, we certainly highlighted the person, team, and product. In one organization, we were able to set up a shared incentive program. An individual developed a unique application that took a feed from the electronic health record and, combined with other medical markers, displayed the output through a wall-mounted display device. As a result, patient falls were reduced by 35%. The vendor who enabled that application inside of the hardware device resold the technology to other organizations. Royalties were collected with each sale. There are many ways to set up success-sharing opportunities, with financial rewards being a strong draw.

Simple recognition is also effective. Most of us have an innate need to feel appreciated and valued. When we delight a customer or develop some sort of breakthrough, it feels good when someone notices and acknowledges the effort. In fact, I believe this is a more powerful motivator than the financial incentive method. At one organization, we would have the person or team who developed something innovative stand before their peers at a monthly town hall and showcase their creation. They would beam in delight. Some might be embarrassed to be recognized in person, so I would send out a firm-wide email highlighting their innovation.

Yet another way to reward and recognize innovation is to hold a contest on a regular basis. We conducted a pitch day where anyone could sign-up and showcase their invention. There were a couple of screening rounds, but ultimately three finalists were selected by our executive team. The opportunity for someone to be able to present before executives is a powerful draw. We saw many proposals that were very impressive. Once the executives selected the finalists, we had each of them do a final pitch before our TEDx audience. We then selected the winner by vote of the TEDx audience. That individual's invention was then funded for full development. One year, we actually hired the person who won to lead our innovation efforts.

At a different organization, we stood up an internal site for innovation. Individuals could submit their invention on the site completing an online submission form. Once the site was closed to submissions, employees throughout the company were invited to vote. All of this was conducted online asynchronously, which made it very convenient. Contributors were recognized, and again the leading vote gatherers were rewarded through modest gifts. Through this process, a couple of the submissions were given the resources required to completely build out, in this case, mobile applications.

We can never discount the human factor in all we do, especially in regards to innovation and the culture required to maximize the opportunities. People want to know that what they do matters for something. When we recognize and reward individuals, it is food for their soul. There are so many different ways to recognize and reward, and the examples I have shared are just ideas that can easily be adopted anyplace. If we are so bold as to ask people to go above and beyond and to innovate, the least we can do is to reciprocate in some way. Your culture will determine what kind of reward and recognition is appropriate, and you can lever up and down as needed. But you must do something. Especially if the innovation fails.

* * *

Fostering Innovations and Supporting Innovators

Inderpal Kohli

How do you help foster a culture of innovation while supporting 24/7 critical care operations? One way is to weave it into the mission of advancing patient care, creating efficiencies and transformation initiatives. Most provider organizations cannot support a dedicated innovation team, and the IT team is consumed with ensuring that the "run and grow" part of the organization executes smoothly, leaving innovation and transformation areas underserved and needing attention.

My recipe for encouraging innovation includes carving out a subset of time and resources aligned with an end goal of increased automation, adding a function, creating efficiency, or transforming an operational function. It does not require a separate funding and resource structure titled innovation and allows our existing team members to participate in an iterative process of designing, creating, and deploying innovative solutions. It motivates our team members and allows them to get out of the cycle of rinse and repeat, which is very common with a growing organization. Growth, which is one of the primary catalysts for an organization's revenue, may sometimes create a monotony of roles for technology professionals because, although every project is different, it is more of the same set of tasks and implementation plans.

An innovative project, initiative, or even a proof of concept can help break that monotony, but it requires vision, thought leadership, bold decisions, and a certain amount of risk-taking. The risk appetite depends on several parameters, including the size and complexity of the organization, financial ability, budget flexibility, and, most importantly, culture. Every innovation project may not be tied to a direct ROI, and it is ok, but you have to determine your organization's tolerance for risk-taking.

I am sharing a few examples of innovation from my recent experiences, where we enabled an advanced clinical function previously unavailable and allowed our team members to gain experience of persistence and sustained effort to bring their efforts to fruition.

Digital Pathology: A Case of Enterprise Imaging and Integrated Diagnosis

The COVID pandemic accelerated the digital transformation of healthcare, both for clinicians and patients, including in the area of clinical imaging. However, one set of clinical imaging was still analog for the most part and could be better incorporated into the enterprise imaging strategy—and that is anatomic pathology.

Anatomic pathology has primarily been based on preparing glass slides of tissue samples for analysis under a microscope. It is essentially a manual process requiring co-location or near-location to process tissue samples and perform diagnosis efficiently. A review of slides with another clinician requires sitting across the table with a multiheaded microscope, collaboration across institutions for research and consults involves couriers and time delays, and teaching and conference presentations involve physical transportation of slides and the setup of microscopes.

The Digital Pathology Landscape

Digital pathology is the process of digitizing glass slides using a whole slide image scanner and a software viewer on a computer monitor to analyze pathology images.

Unlike digital radiology, digital pathology still has a manual component in that tissues on a glass slide need to be scanned for digitization. Another difference is that radiology has a standardized file format called DICOM instead of the multiple formats used in pathology.

That said, digitizing slides and image availability in the enterprise system offers many benefits, including enhanced multidisciplinary collaboration, research, and teaching to name a few. While the practice of using digital images and workflows for pathology diagnosis is well established in Europe and other parts of the world, we are now seeing digital pathology take off worldwide due to the advent of faster whole slide imaging. In the United States, the Food and Drug Administration (FDA) also approved software and scanner combinations for primary diagnosis, which acts as a catalyst for the increased digitization.

Benefits of Digital Pathology at the Enterprise Level

Digital access to all patient images will lead to more informed analysis and a better clinical diagnosis. Institutions that plan to implement an integrated diagnostics setup will benefit from the correlation of pathology and radiology images. Integrated diagnostics enables all imaging to be housed either in a single PACS system or within the enterprise EMR. At HSS, we utilize Sectra PACS for both radiology and pathology images.

A digital capability also decouples the pathologist's physical location from the histology lab, especially when providing pathology services to remote and rural areas. Furthermore, enabling pathology services beyond physical proximity is helpful as the overall number of pathologists in the United States decreases over time. (In 2019, *JAMA* published a study reporting that the number of pathologists in the United States has decreased by 18% between 2007 and 2017, while the workload has increased by 42% per pathologist.) It also enables efficient collaboration across institutions and industries for research and multisite studies, eliminating the physical transfer of slides.

The software solution also offers image analysis tools to create efficiency and reduce manual tasks such as cell counting or percentage cancer tissue calculation, providing a standardized image analysis across the institution, thus eliminating the discrepancy between diagnoses.

How to Optimize Infrastructure and Integration

The software solution and infrastructure should support server-side rendering of large images, the lack of which creates a suboptimal experience for the pathologist. Plan for significant storage capacity as the size of pathology images can range between 500 MB–2 GB, dependent on the magnification.

Consider integrating pathology imaging as part of enterprise imaging infrastructure to realize maximum benefits. The availability of pathology imaging from the same portal/viewer as other imaging eliminates the need for additional training and helps with clinical adoption. The solution should support the most prevalent file formats (svs, ndpi, etc.) while waiting for the DICOM standard to gain traction in pathology imaging.

If a single-vendor solution is an option, institutions can benefit from extending the existing PACS infrastructure, knowledge, and administration for digital pathology. A LIS/EMR integrated solution would not require a separate tool outside of the standard workflow, eliminating a significant change for the pathologist.

The Value of Integrated Diagnostics

We took a leadership role in partnering with our digital pathology and enterprise imaging partners, Sectra and Epic EMR, utilizing the Leica Scanner to cut the trail for the integration path for half of the nation's healthcare systems. We did comprehensive development work that other organizations on Epic can easily leverage to enable integrated workflow.

Intending to enable an efficient diagnostic process while maintaining a familiar workflow for pathologists, we invested in developing the unique integration between Sectra and Epic to allow image launching from the pathologist worklist. A pilot project to scan images of particular clinical interest into our PACS and share them with other clinicians was implemented while waiting for the FDA clearance. It served two purposes: It allowed us to optimize workflows and system utilization, and it provided encouragement for our teams who have been working on the development project for months, if not years.

The pandemic accelerated the FDA approval timeline of this solution, and thanks to a successful pilot, we deployed an FDA-approved solution in a very short time.

This was a genuinely innovative pioneer solution that will change the way pathologists perform their roles and collaborate with other clinicians. The best way to reward and encourage all who were instrumental in this achievement was to bring them all to the front. I did this by providing coverage through press releases and internal communications at the department and enterprise levels and celebrating the launch. The success of one innovation project led to renewed interest in others, which is what is needed to create a pipeline of solutions, excitement in our teams, and continued support from the executive leadership.

What Does the Future Hold?

The technological evolution of digital pathology is just getting started, and one noticeable advancement includes the use of machine learning for image analysis. Deep learning or AI algorithms can now analyze complex visual features in image data, assisting pathologists by automating the tedious tasks of slide image analysis.

With the advances in image analysis and the ability to provide quantitative information about biological samples, digital pathology will help pathologists make more accurate and consistent diagnoses, leading to improved patient outcomes.

Recognition and Rearwards: The Fruits of Our Labor

While the immense satisfaction of building and deploying a pioneer technology solution is a reward in itself for all involved, I ensured that the launch was widely celebrated, and our team members were recognized for their innovative and formidable work. We partnered with our internal communications and vendor partners to promote this achievement in the print and online media, including interviews with our team members.

We also took this on the road and presented at industry forums, grand rounds, conferences, and user meetings. It helps create an encouraging environment by promoting innovative projects and thereby recognizing innovators for their perseverance and hard work. All of us in healthcare are open to collaborating and sharing with a common goal of advancing patient care and helping all patients. The trail we cut with this-first-in-the-country integration allows other Epic organizations to reach the same goal in a much shorter time. Our team members and I find immense satisfaction in being the enablers and consider this our greatest reward for making a difference in better patient care.

Another example of the recent implementation of innovative technology solutions is a unified communications platform.

Zoom as a Unified Communications Platform

Another lesson from the recent pandemic is the emergence of the need for unified communication across the enterprise and collaborators. The seamless alignment across teams for consistent communication and collaboration is the key to productive teams and organizations. In healthcare, this collaboration takes a special meaning in allowing physicians, specialists, nurses, and other care providers to collaborate for better patient care. It is no secret that collaborative and informed decision-making results in better patient care, which is why interdisciplinary rounding is a standard process. A unified communications platform enables a virtual or hybrid interdisciplinary care model by enabling seamless connectivity across teams, regardless of location. This is not necessarily only to engage distributed remote teams but also in-house clinicians who do not have a desk job and are consistently on the move.

An Opportunity for Innovation

An aging and soon-to-be out-of-support PBX system presented an opportunity to dream of an innovative approach with Zoom. With this forward-thinking approach, we became the first hospital on a Zoom unified communications platform and created an innovation endeavor for our teams, while meeting the business need. Besides gaining experience with the newest technology, the motivation is to enhance the mission of patient care with provider efficiency and ease of communication. The PBX platform replacement with IP-based telecom solutions is a traditional route for most organizations. We seized on the opportunity to develop a solution to connect clinicians no matter where they are via seamless mobile, desk, and softphone; the ability to text/chat; and quick access to the care team. It requires bold leadership, risks, and being on the unknown path with a commitment to stand behind the team. Recognizing and letting our team know of the challenges of the unexplored path goes a long way in encouraging the team to be courageous and think beyond the traditional norms, which are critical for the success of all pioneer initiatives.

Benefits of Zoom Unified Communication Platform

Omnichannel communication: With Zoom, we implemented a full-featured cloud telephony solution that is easy to use and administer while providing a modern integrated user experience across Zoom meetings, desk phones, softphones, and mobile phones. The enriching experience allows transfer and portability of communication between various channels and easy-to-use communication tools to streamline interactions with colleagues and patients. Clinicians are always on the move between hospital units and clinics, and the unified communication platform can allow them to interact with the care team via remote video conferencing, messaging, and phone calls. The call flip feature enables them to seamlessly transfer calls between devices with one button, simply by putting the call on hold and picking it up on another device.

Integrated Telehealth: The exponential utilization of hastily put-together telehealth solutions during the pandemic accelerated the need and development of HIPAA-compliant platforms that are easy to interact with patients and clinicians. Zoom platform provides a streamlined experience for patients and clinicians, utilizing a secure platform for healthcare communications that meets HIPAA and SOC-2 compliance. We also implemented telehealth via Zoom during the COVID pandemic, which met its compliance requirement but is not the most integrated and streamlined experience for our clinicians and patients. We challenged our teams, including our vendor partners, to improve the workflows, experience, and efficiency of patients and clinicians. Our ability to utilize a unified communication platform by Zoom allows clinicians to interact with a familiar set of tools and benefit from its many features.

Manage patient communication and record interactions: Recording a patient interaction on the fly is as easy as pressing a button to record the call, but in a secure and compliant platform, versus your phone's voice memo. It creates both the audio file and transcription for future review. Voicemails can be delivered straight to the email inbox as an audio file and transcription. This includes shared voicemails and voicemails to delegated users, so a receptionist could receive a patient voicemail sent to the main line or specific provider. You can even export that voicemail to a patient's electronic medical record.

Manage patient communications and protect provider privacy: Zoom Phone allows you to easily manage your outbound caller ID number with just one click, protecting the privacy of doctors and nurses by allowing them to present their office number instead of their direct line or cell. Also, your receptionist or office manager can streamline patient interactions by changing their caller ID to a specific department number, scheduling phone number, or central line when calling patients.

A single client across all devices: Whether your staff members and providers use an Apple or Android, Mac or PC, Zoom Phone works with a single client across your desktop and mobile devices. It's the same app used for Zoom Meetings and Zoom Team Chat, meaning your users are already familiar with it.

Our experience with Zoom's multi-purpose platform offers ease of use and helps healthcare professionals and patients enable programs and safeguard PHI. This allows healthcare organizations to reach across the virtual care continuum, including healthcare administration, medical education, and telemedicine. It offers an integrated workflow from an EMR to mimic provider and patient workflow or, if opted for, a traditional Zoom meeting experience.

Over the years, the EMR implementations intended to increase data transparency and promote efficiency might have increased the "busy work," resulting in fewer doctor–patient interactions. Fortunately, awareness of that lost aspect of care is part of the new technology design where ambient and intelligent automation takes root in future product blueprints. There is an increased focus on maintaining doctor–patient communication, which now transcends beyond the traditional face-to-face model. Technology is helping evolve these bonds to digital interactions, especially with changing demographics and busier lifestyles. While it may have suffered in the past, the future designs are for an engaged patient and an interactive clinician aided by technology and ease of use to strengthen these bonds further.

Recognition and Rewards: The Fruits of Our Labor

From my previous experience, we recognized the need to consistently encourage and reward our team, both for being the innovators and for deploying and supporting such solutions in a fast-paced, acute care environment. In this case, other than the traditional methods of rewarding the team, I found that standing behind them at all times was the most important reward and recognition. The critical post go-live period, where users and technology are still adapting to each other, requires unwavering leadership support and small snippets of recognition where possible. We also published and celebrated our pioneer unified communication initiative in both state and local print and online media. We recorded a recognition video demonstrating our technology's ease and use, which was shared at various forums. Both clinicians and team members participated in the video. I am playing the role of the team's spokesperson, and promoting our innovative solutions is one of the few ways we reward our innovators.

* * *

Recognizing and Rewarding Innovation

Michael Fey

Overview

Innovation in every industry is as challenging as it is elusive, part science, part art and part magic. As the world's leading cyber security company, innovation is core to Symantec's mission, and in our experience, the best way to spark innovation is to create a culture where it is a fundamental part of everyone's day-to-day work. We never want our employees saying to themselves: "Today, I need to spend the next six hours on my assigned tasks, and then I'm going to spend two hours on innovation." Rather, innovation should simply be built into the fabric of each employee's daily activities. This is, of course, easier said than done and requires ongoing encouragement, recognition and rewards for developing breakthrough ideas. Following are a variety of programs and approaches we've implemented in our company that highlight to employees the premium we place on innovating across every aspect of our business.

Innovation Days

We've had a lot of success running regular "Innovation Days" in our organization. These events occur at different sites, where employees (typically engineers, in the case of our company) organize into teams of two to four people with the goal of building a quick prototype of a new idea over the course of one to two days. Each team then presents their prototype/demo to the entire group, and the "winners" are chosen by various mechanisms (e.g., a vote of all participants, a panel of judges, etc.). In our experience, the ideas generated at these events are often innovations that offer great benefit to customers and can be commercialized relatively quickly. In fact, at our initial Innovation Day events, we would often find ourselves asking engineers, "Why didn't you propose your great idea as part of your regular day-to-day job, why did you wait until this innovation event to propose your idea?" The response was often, "I assumed that I needed to do my assigned tasks before proposing some new idea, and my assigned tasks take up all my time." From these early experiences, we've revised our product development processes to help ensure engineers get the encouragement and the opportunity to propose innovations as part of their everyday jobs. Based on their feedback, these changes have made our engineers feel more appreciated and have helped create a steady stream of innovations. And our Innovation Days continue to be a fun and popular way for our engineers to prototype their innovative ideas, helping to create a strong pipeline of potential new products and technologies.

Test Drives

We have also found that a great way to inspire innovation and energize our engineers is to give them exposure to work occurring in other teams that may spark new ideas. To help drive cross-pollination, we have instituted a series of "Test Drive" events. At each Test Drive session, experts from one specific area of our business (e.g., software developers working on a particular technology) run a training/deep dive so that other employees can learn more about the area. The primary goals of these sessions are to cross-fertilize ideas between teams, to expose engineers to other areas of the company that may be of career interest to them in the future and to give the "trainers" a chance to highlight the interesting innovations and new capabilities they've implemented in their areas of focus. At the end of the training, the "students" are given a series of technical problems to work on and to reinforce what they've learned. We also keep an online leaderboard that's a fun way of showing the progress of everyone working on these exercises. We've found that these sessions often lead to teams thinking about problems in new ways and finding new collaborators, leading to new ideas that would have been much less likely to occur without cross-team collaboration.

Patent Program

Protecting intellectual property is an important part of commercializing innovation. An effective patent program not only encourages employees to protect the IP developed by their novel ideas, it can also help spur break-through innovation. Our patent program includes several components that have proven to be very effective.

First, at our company, any employee can submit a relatively simple invention disclosure to our internal Patent Committee, outlining the details of their innovation. Once a disclosure is accepted by our Committee, a patent attorney is assigned to work with the employee to formally file a patent application with the U.S. Patent Office—the attorney does the bulk of the work in this process and takes minimal time on the employee's part. Employees receive awards once their initial disclosures are accepted by the internal Patent Committee and are given additional awards once their patents have been filed and/or granted. In the past, these awards have included monetary rewards, patent jackets, plaques, or other recognitions. These awards serve as incentives for employees to protect intellectual property and also provide recognition and encouragement to pursue breakthrough ideas.

To ensure the success of a patent program, it is very important to provide appropriate training for employees. This training should cover how employees can identify that a specific portion of their work is potentially patentable and how to easily and quickly communicate the essence of the breakthrough idea in their initial disclosure to the internal Patent Committee. Ideally, the process required for employees to protect the IP generated by their innovations through patenting should require very little additional effort beyond conceiving the innovations and sharing the relevant details with a patent attorney.

Recognition Awards and Events

We have implemented a variety of different mechanisms to recognize innovators in our organization. These approaches have a broad range of costs and occur at different frequencies, giving us a diverse set of opportunities to drive our culture of innovation. These recognition awards and events are listed in order roughly from lowest to highest cost:

1. **Thank You Notes from Senior Executives**

   We have found great benefit in having senior executives send thank you notes (typically via email) to specific employees expressing appreciation for contributions to the success of our business, including for new innovations. These notes provide employees with very personal recognition of their work, and we've found that the recipients are very pleased to receive them directly from senior executives.

2. **Recognition at All Hands**

   Several of our executives have recognition programs for their own organizations, allowing them to highlight and acknowledge great work from specific employees at all-hands meetings. This gives each executive the chance to highlight innovation or other contributions in front of a large audience, and we have seen the employees being recognized take great pride in receiving congratulations in front of their peers/co-workers. These awards can also be accompanied by monetary bonuses (as permitted by local law)—we've found that even relatively small monetary bonuses are much valued by the recipients as these add some weight to the public recognition.

3. **Employee-Nominated Awards**

   Internal programs in which any employee can reward any other employee for their work in delivering new innovations or making other successful contributions to the company are also effective. These awards can also have a cash bonus component. We've found that our employees greatly appreciate being empowered to directly reward the great work and innovation of their colleagues.

4. **Innovation of the Year Awards**

   Each year senior members of our company's Patent Committee propose a short list of those patents filed in the prior year that were found to be the most innovative and valuable to the company. Our CEO then chooses from this final list to recognize the Innovation of the Year, which entitles the named inventor(s) to special recognition and monetary awards above and beyond those that are offered for other patents filed under the patent program. In addition to receiving special recognition from the CEO, the inventor(s) and their winning inventions are highlighted in a special article on the company's intranet.

5. **Recognition Events**

   Another effective program to consider is holding regular (ideally annual) innovation recognition/award events. In our company, we've used such events to recognize employees who have been successful at filing patents, as well as highlighting specific innovations/innovators that have made a significant impact to the business (e.g., "Innovation of the Year"). Another example is to recognize the best innovation from a more junior employee (e.g., an employee who has only been at the company for a limited period of time, or who is below a certain level of seniority). By recognizing innovations from more junior employees, as well as the most important innovations from all employees, we help inspire a new generation of talented people at the company to innovate. It has been our experience that these events play as important a role in recognizing and encouraging innovation as the direct financial benefits that employees receive for patents that we file. If holding a central event for all innovators is prohibitive from a cost perspective, we recommend doing smaller regional events, attended by senior executives, and/or recognizing top innovations on the internal company website.

Innovation Newsletters

Regular innovation newsletters have proven to be very effective at our company for keeping innovation top-of-mind as well as highlighting exciting new products/technologies that have recently shipped, cool prototypes and early-stage research that we're developing in our labs, upcoming events (e.g., Innovation Days, patent training, etc.), interesting curated online information around innovation, etc. These newsletters serve to highlight the innovators and their work across the company, reinforcing the importance and culture of innovation that we seek to drive throughout our organization.

Fellow and Distinguished Engineer Program

We have two engineering titles reserved for our top innovators at the company, Fellow and Distinguished Engineer (both relatively common designations in the industry). These titles specifically recognize engineers who have made breakthrough innovations of importance to our products, our company and the industry. We also have more traditional engineering titles (e.g., architect) intended to recognize engineers who have designed and provided technical leadership and implementation on very successful products in the market. We have found that the Fellow and Distinguished Engineer roles have provided a great career path and meaningful recognition for our top innovators.

Learning from Failure—Black Box Thinking

One of the most powerful ways that we've found to recognize and inspire innovation comes from a very unlikely source—failure. All too often, when something goes wrong inside a company, the response is to try to isolate the explicit cause of the failure, fix it quickly and move on (or worse, try to cover it up so nobody notices). By contrast, we've embraced the idea of "Black Box Thinking," a name coined by Matthew Syed and taken from the airline industry. Commercial airlines have steadily improved safety over time by objectively analyzing data from their on-board black box after any accident and then implementing steps to address any discovered issues. In our organization, we examine failures by looking at the entire end-to-end process that failed—we don't simply isolate our investigations to the point of failure; rather, we try to discover systemic improvements we can make to the entire system. Numerous innovations and improvements have come from such black box analyses among our teams, and this approach has helped create a culture where continuous innovation and improvement are recognized and rewarded.

Summary

These are a few of many possible mechanisms to encourage and inspire innovation in your organization. Every time you run such an event or activity, it provides another opportunity to highlight that innovation is not an activity done by a separate group or overseen by a special department. Rather, innovation is the job of every single employee, every single day.

* * *

Recognize and Reward

Pamela Arora

Recognize and Reward—The AAMI Awards

Upheaval, transformation, and revolution are all words we commonly associate with innovation. Every industry sector strives to implement new ideas or improve its methods, but it is without hesitation that I would say the healthcare industry epitomizes that goal.

Certainly, it was witnessed as the global medical community alleviated burdens from the COVID-19 pandemic by developing diagnostics and vaccines to serve the global populace. However, innovation also plays out in the far, unseen corners of hospitals, research institutions, and other health services. Every day, trained professionals following industry standards ensure our healthcare remains safe and operational. While they all are heroes, it's sometimes only possible to recognize a select few that go above and beyond their call of duty. And, while award recipients deserve recognition, it's also important to acknowledge the individuals who nominate their peers for their transformative contributions. These practitioners help feed the innovation pipeline.

The Association for the Advancement of Medical Instrumentation (AAMI) has been spotlighting these exceptional specialists for nearly half a century. AAMI is pleased to honor individual achievements in healthcare and the contributions our members make to the association and the wider world. This recognition helps motivate and inspire the healthcare community to educate, collaborate, and innovate.

AAMI is a diverse community of more than 10,000 member volunteers united by a mission of developing, managing, and using safe and effective health technology. These professionals serve the global community by developing superior ethical healthcare standards, fostering a trusting and inclusive health environment, and promoting effective technology advances that improve human health.

Our members are on the front lines of healthcare. These individuals leverage their unique resources to solve problems, advance industry or professional performance, kick-start innovation, and improve conditions around the world. They react to current demands and predict future needs. In short, they help to save lives every day.

To recognize the outstanding accomplishments of these exceptional professionals, AAMI developed an awards program to spotlight leaders from cross-cutting sections of the healthcare community.

AAMI initiated its awards program in 1975 with the launch of the Laufman-Greatbatch Award. Named after two pioneers in the field—Harold Laufman, MD, and Wilson Greatbatch, PhD—this highly regarded award honors an individual who has made a unique and significant contribution to the advancement of healthcare technology and systems, service, patient care, or patient safety. Considered by many as AAMI's most prestigious award, its success resulted in the continual growth of the awards program. AAMI extended its award offerings as the industry evolved. Now, each year,

AAMI recognizes approximately 20 innovators and health technology leaders whose efforts have moved the industry forward and provides those recipients with awards, grants, and scholarships.

AAMI's award offerings have evolved over the years. As AAMI's mission expanded to meet the needs of our members, we also provided platforms to celebrate visionary leaders within those professional settings. While all the awards have similar underpinnings that honor achievements in healthcare technology, each award is unique and recognizes distinct contributions. Recognition programs range from celebrating young professionals and the achievements of biomedical equipment technicians to improving global human conditions. In 2021, AAMI launched a new award that recognizes an individual who has demonstrated the vision and leadership necessary to solve some of the industry's most critical medical device cybersecurity challenges.

AAMI and the AAMI Foundation currently offer the following awards:

- The **AAMI Foundation's Laufman-Greatbatch Award** honors an individual or group that has made a unique and significant contribution to the advancement of healthcare technology and systems, service, patient care, or patient safety.
- The **AAMI Foundation and ACCE's Robert L. Morris Humanitarian Award** recognizes individuals or organizations whose humanitarian efforts have applied healthcare technology to improving global human conditions.
- The **AAMI Foundation and Institute for Technology in Health Care Clinical Solution Award** honors a healthcare technology professional (individual or group) that has applied innovative clinical engineering practices or principles to solve one or more significant clinical patient care problems or challenges facing a patient population, community, or group.
- The **AAMI Foundation and TRIMEDX John D. Hughes Iconoclast Award** recognizes an individual who pushes the boundaries of the healthcare technology management profession and demonstrates individual excellence, achievement, and leadership.
- The **AAMI Patient Safety Award** recognizes outstanding achievements by healthcare professionals who have made a significant advancement toward the improvement of patient safety.
- AAMI's **HTM Leadership Award** recognizes individual excellence, achievement, and leadership in the healthcare technology management profession.

- **AAMI and GE Healthcare's BMET of the Year Award** is given to a biomedical equipment technician to recognize individual dedication, achievement, and excellence in the field of healthcare technology management.
- AAMI's **Young Professional Award** is presented annually to a professional under the age of 40 who exhibits exemplary professional accomplishments and a commitment to the healthcare profession.
- The **Spirit of AAMI Award** recognizes the outstanding contributions of an AAMI member in volunteer efforts within the association.
- AAMI's **HTM Association of the Year Award** recognizes an AAMI healthcare technology management association that distinguishes itself during the course of the year through outstanding society operations and meetings as well as a commitment to elevating the healthcare technology management profession at the local level.
- The **AAMI and Medcrypt Cybersecurity Visionary Award** recognizes an individual who has demonstrated the vision and leadership necessary to solve some of the industry's most critical medical device cybersecurity challenges.

Awardees represent the best and brightest in the industry and are often nominated by their industry peers. Most awards include monetary prizes, and all award winners receive a plaque commemorating their achievements. Further, awardees are celebrated by the broader AAMI community at its annual conference and exposition, the AAMI eXchange.

Recognition Done Right

Although AAMI awards are typically conferred in the summer at our annual conference, the awards process spans year-round. The competitive application process initiates in the fall with a call for submissions and nominations and remains open until January. Staff promotes the award nomination process and amplifies its marketing campaign via promotional information on the AAMI website, newsletters, email campaigns, and advertisements in key publications. In most cases, individuals are nominated by their colleagues, friends, or others based on their qualifications and background. Individuals may self-nominate for certain awards. In either case, they are required to submit a formal application, written statement addressing how the nominee has met the qualification criteria, letters of recommendation, and a curriculum vitae. Some AAMI awards recognize specific and recent accomplishments, while others recognize lifetime achievement or volunteer activity. Nominators should be careful to distinguish this in their submission. AAMI membership is not a requirement for all award categories; however, because contributions to the healthcare community are important criteria in the selection of award recipients, AAMI service or participation may be a discriminator between otherwise equal candidates.

While AAMI is receiving applications, it assembles the AAMI Awards Committee. The committee reviews and selects final recipients. It also annually evaluates each awards' criteria for adequacy, redundancy, clarity, and for applicability to AAMI's membership. In order to provide a balanced and fair review of the disparate slate of global candidates, this prestigious committee comprises distinguished healthcare professionals with diverse background experience. The opportunity to apply to serve on the Awards Committee is open to all AAMI members. While the committee is managed by AAMI staff, it ranks highly important to the organization's mission and, therefore, reports directly to the AAMI Board Executive Committee. Further, the Chair of the Awards Committee is selected by the AAMI President, subject to Executive Committee review.

In early winter, AAMI assembles all applications, categorizes them into respective award buckets, and disseminates information to the Awards Committee. The group assesses the quality of the nominations and uses a grading system developed by AAMI staff to ensure each application is fairly evaluated. Grading criteria varies per award and includes four types of metrics: Sufficient examples of candidate's relevant work, quality and number of letters of recommendation, comprehensive CV or resume submission, and strength of personal statement. The grades given by each committee member are compiled by AAMI staff and used as a starting point for discussion on the annual selection call. Final recipient selections are determined by a majority vote during that call.

After the committee deliberates on the applications and votes on the recipients, staff notify the winners about this special achievement. Winners are invited to share their stories on AAMI.org and across AAMI's social media channels. Press releases with remarks from award recipients are issued. Comprehensive articles are published in AAMI newsletters and disseminated to tens of thousands of readers. A celebratory video is produced and posted on the AAMI.org home page, sent to AAMI members, showcased on social media, and shown at the AAMI eXchange. Finally, physical awards are conferred by AAMI's President and AAMI Board Chair during plenary sessions of the annual AAMI eXchange conference and exposition.

The most noteworthy part of this entire process is how it gives our awardees one more platform to share those ideas driving revolution, transformation, and yes, even upheaval in healthcare. When exceptional thought leaders and innovators grace our professions, recognition is due—there is no doubt—and AAMI strives to make sure it is done right. However, when the time for final recognition is upon them, you'll never see an AAMI awardee just basking in the limelight. They're already finding collaborators, carefully considering next steps, and seeking ways to further improve themselves or their field. It becomes clear then, after an AAMI Awards ceremony when the crowds disperse and awardees mingle, that *our* reward is the assurance that this will not be healthcare's last opportunity to *recognize* and *reward*—far from it.

* * *

An All-Important Part of Sustaining Innovation in Your Company

Jonathan Scholl

With apologies to Hamlet and Shakespeare, if we are speaking about innovation in today's world, "to be or not to be" is not the question.

The question is "how?"

How do you keep the flame of innovation alive at all levels and in many different locations inside a big organization—knowing that to do so is critical to the enduring success of the enterprise?

Or, to look at the same question from another angle, how do you maximize innovation without sacrificing the discipline needed to take care of business and satisfy the customer in the here and now of today?

Managing innovation is a difficult, but by no means impossible, balancing act. At Google, they say, "Take Friday and invent something." At Leidos, we may be a bit more structured in our approach to innovation. To inject the necessary level of practicality into our thinking, we define innovation as "the implementation of new ideas with business impact." We look for new and valuable ways to solve difficult problems for our customers. In our contract R&D work for a variety of governmental and commercial customers, we aim to develop products or systems that they will be eager to purchase in expectation of getting an excellent, or even an exceptional, return on their investment.

Probably everyone who is reading this book would agree: Innovation isn't hierarchical; good ideas can come from anywhere. However, to be successful in promoting a culture of innovation, your organization has to have the optics to see it. There should also be an assortment of public mechanisms for celebrating and rewarding innovation.

With that in mind, I will begin with a short self-introduction, as Leidos is not exactly a household name (and, to answer the first, most frequently asked question, the company name is a creative shortening of the word *kaleidoscope*, which itself comes from the Greek *kalos*, beautiful; *eidos*, forms; and *skopien*, to see).

Establishing a Culture of Innovation Attracts Employees Whose Reward Is the Work Itself

I've long said that culture is the only sustainable competitive advantage that a company can create. Products and service offerings, financial positions, and market positions can be replicated with some effort. But replicating a culture is tougher. Look only to the automotive industry, where Detroit's capabilities in quality management took decades to replicate its foreign competition. I recall a story about a factory worker in Japan leaving the plant after their shift and, as they were walking out, straightening windshield wiper blades so they were "just so" on the new cars. Culture drives meaning, purpose, and a willingness to invest beyond the norm.

So I offer a brief review of the culture of Leidos, best illustrated by a brief history, so that you might understand the length of time and leadership attention needed in establishing a culture where innovation thrives and is its own reward.

Leidos at Nearly 50 Years—A Short History of Who We Are and How We Got to Where We Are Today

It's no guarantee of future success, but it certainly helps if the founders of the well-established company you work for imbued their creation with some of the best aspects of their own DNA—as seen today not just in the technical smarts of the people but also in the retention of a good deal of the original character and purpose of the company. At Leidos, we think we have been very fortunate in both regards. The lesson for innovators? It starts with culture. Leaders have to live—and reward and advance—innovation as part of their own ethos.

J. Robert Beyster, the founder, was both a brilliant scientist and an unselfish (I could also say *enlightened*) manager. Born in 1924 to parents of modest means living in Detroit, Michigan, Beyster grew up during the Great Depression. Shortly after Pearl Harbor, at the age of 18, he enlisted in the U.S. Navy and served on a destroyer. Following the war, he went to college and on from there to earn a PhD in nuclear physics at the University of Michigan. That was in 1950, when he was 26.

Bear in mind: This was the golden age of physics—and of nuclear physics most especially. The biggest names in science at this time were Albert Einstein, Otto Hahn, Enrico Fermi, William Teller, and other luminaries in nuclear physics. They bestrode the scientific world like the greatest of giants.

In the 1950s and first part of the 1960s, Dr. Beyster worked for a company called General Atomics. This was a hot time for start-ups and stocks related to nuclear energy—similar to, if not so numerous as we have seen in more recent time with the booms in dot.com and biotech start-ups. Our founder worried that too many of these young companies were dedicated to the personal enrichment and glorification of small numbers of people. He wanted to create a different sort of company—one in which *all* of the people working for the company would do more than just share in the financial rewards of ownership; they would see themselves as stakeholders and principals in an exciting business. In a handwritten organizational plan for a new company, he called for employee ownership as a hallmark of the company.

In 1969, at the age of 45, with a wife and three kids to support, Dr. Beyster used the family home as collateral to secure a bank loan and invested $50,000 to fulfill his dream—launching a company first known as Science Applications Incorporated (SAI), later to become Leidos. SAI earned a grand total of $20,000 in 1969. Dr. Beyster rubbed his eyes in disbelief when he did the accounting. "After a year, a surprising thing happened," he exclaimed. "We made a profit!"

If you look at our company timeline over the next several decades, you see a growing involvement in big and important events, all of them opportunities captured by a workforce driven to bring science into applied use in serving our customers. Among other things, there were these achievements:

■ SAI played a key role in orchestrating the clean-up operation after the partial meltdown of a reactor at the Three-Mile Island nuclear station in Pennsylvania in 1979.

- In 1983, President Ronald Reagan announced the Strategic Defense Initiative (SDI), a missile defense program to defend the country from a Soviet nuclear attack. Two years later, the Pentagon awarded SAI a $5 million contract to research how SDI's anti-missile system should be designed. DoD chose us as a key integrator in the program that was dubbed "Star Wars."
- In 1986, SAIC (as SAI became known after the name was altered to Science Applications International Corporation) collaborated in the design and building of an unmanned aerial vehicle that prefigured later generations of drone aircraft.

Even so, SAIC remained almost completely under the radar screen in terms of public recognition until 1987—when we won our first 15 minutes of fame for what was purely a fun and patriotic exercise in innovation. SAIC co-engineered *Stars and Stripes*, the America's Cup winner in that year. In defeating the Australian defender *Kookaburra* four races to nil, she brought the cup back to the USA in high style. With our help, *Stars and Stripes* set a new standard in hydrodynamic design.

After 34 years as CEO, Dr. Beyster retired in 2003 at the age of 79. Two years later, the Board of Directors voted to take the employee-owned company public. Was this a sudden reversal in the original character and purpose of the company?

Not at all.

Culture is persistent: The company has not lost the spirit of being a place where people can continue to think and act as owners—or as self-propelled entrepreneurs in a high-tech, high-stakes business.

Innovation Continues . . .

- Thirty years ago, Leidos built the first electronic health record for the DoD. That's something we are replacing today. Over all that time, we have managed the health records of all of active-duty and retired military personnel and their dependents. The current number of beneficiaries in this database is close to 10 million people. We also have more than 70 health clinics around the country serving veterans and providing access to a network of more than 12,000 physicians and caregivers.
- We also developed a highly sophisticated platform (called LEAF for Leidos Enterprise Application Framework) that we leveraged to develop solutions that the U.S. Air Force uses at its operational centers around the world. Right now we are making a big effort to extend LEAF's reach

through similar interoperable platforms to serve hospitals and other providers of healthcare. Our goal is to eliminate waste, reduce costs, and help optimize the delivery of care for millions of people.

■ Finally, we are doing pioneering and exciting work in the field of biomedicine. At Leidos, we created the first Zika antiviral vaccine based on the human genome. With the scientists at Leidos Biomedical Research (which manages the NCI), we have helped advance curative therapies for certain cancers—specifically, a certain type of neuroblastoma—using big-data analytics combined with cutting-edge advances in using the human body's immune system to destroy invading cancer cells.

Gabe Gutierrez, one of a prestigious group of Technical Fellows at Leidos, leads a small unit that has been working closely with the NIH and DoD. As a Leidos employee since 2007, he was pleasantly surprised to find that there is far more teamwork and camaraderie at Leidos than there was in the academic world that he left behind. Still more, he found it was easier to think big—in assembling the resources to focus on major problems.

When Gabe, who has a PhD in genetics, speaks about the now-gathering immunotherapy revolution, it is impossible not to share his sense of wonderment and excitement.

We may be coming to the end of an era in medicine when it was common practice to inject poison into people's bodies in the hope that it would kill the cancer before it killed the victims of the disease. Now we are learning how to create and deploy new molecules that alert the body's immune system to the presence of invading cancer cells and enable it to destroy them and save the patient.

Leidos Health—Reaching from Lab Bench to Bedside through Innovation

The health business within Leidos has the highest concentration of PhDs in the company. So we have a lot of very talented people. But how do we motivate them? How do we recognize their contributions to the enterprise? And how do we reward them?

As I see it, there are four key elements. We give them the following:

1. Opportunities to do interesting work.
2. Time and resources, for the many and not just the few, to pursue their passion.

3. Recognition and accolades woven into the social fabric of the business.
4. Rewards, incentives, and career progression that works for creative individuals who prefer product innovation to management or sales.

Let me address each of those points.

First, there is no greater gift that you can give to inventors and other creative people than plenty of opportunity to do interesting work. You can start with the worst problem your customer is having in some area (one example: Doing what Gabe does trying to stop the spread of certain cancers) and turn it over to your most innovative and creative people to come up with a solution. They will love the challenge. That's what they live for.

We are lucky at Leidos to have hundreds of assignments every year with our customers—thousands over the span of years—and therefore, we have the ability to rotate people and give them a variety of work and opportunities to work in areas around their passion. To accomplish this, we have deliberate structures that help our people manage their careers, see the new work that is emerging, and voice their preference in work assignments—preference that cannot always be fulfilled but often can.

Second, and this is like the Google culture mentioned previously, we don't want to have too narrow a focus on the top guns. We also want to capture the creativity of many other people across the enterprise. We encourage our employees, particularly the scientists and engineers, to propose innovation projects every year. Their submissions are reviewed by senior panels of respected innovators and leaders. Every year, we make a number of awards—with projects receiving $50,000 in funding to pay for their time and materials in pursuit of their ideas. A year later, senior executives review the results and express their personal gratitude to individuals who have done an outstanding job of exceeding customer expectations and enhancing the company's reputation for innovation and technical excellence. The selection, and the recognition for the work, reaches the highest levels of our organization. Our executive leadership team and board of directors often review and see the work that our people do, reinforcing the culture.

Third, many companies do a great job at finding ways to motivate high-potential people (or "high pots," as they are sometimes called) in sales, finance, and executive management, but very few give equal attention to inventors and talent in technical positions. That is a bad mistake. We believe that recognition and accolades for innovation should be every bit as common and command the same recognition and respect as "manager of the year" or "salesperson of the year." The high-potential employees with technical expertise and talent should be singled-out and put up on something of a pedestal for their achievements in innovation, no less than the up-and-coming stars in other important company functions. Without going into all the details here, we have done that at Leidos in showing our appreciation of and respect for two levels of achievement: One for "technical fellows" (fewer than .05% of Leidos technical staff) and another for "senior architects," mostly reserved for a few holistic thinkers with the multi-dimensional gift of seeing things through in complex programs from initial design through development and the resolution of problems that crop up in early implementation or production. We celebrate the dedication and achievements of these same people in a multitude of other ways as well, including them in big annual events that we call "technical showcases," celebrating their achievements in internal publications, and encouraging them to attend speaking engagements and international forums at the company's expense.

Finally, make sure you have a career path that does not lead to a "Dead End" sign for the technically minded people who love invention and innovation. Do not force them to go into management. Pay them well, and allow them to continue to flourish by staying in the laboratory and continuing to be creative. At Leidos, we have a technical career track that lifts the most senior, accomplished innovators to a pay rank commensurate with their contributions to the firm, and our best innovators and scientists are paid like management.

It's not really about the money. I've found that the best innovators make their contributions for the love of the work that they are doing. But employees—even the inventors and innovators—don't like to be ignored or taken for granted when they know that they are making a great contribution to the well-being of the company or organization. There has to be a base-level expectation for your technically minded employees that they are being recognized and treated as the integral part of the company that they are.

In short, celebrate the success of some of your smartest people and make sure they know you're passionate about them—almost as passionate as they are about the work they're doing!

Chapter 8

Co-Create Solutions

Appreciate the complexity of attention that innovation requires,
and expose the organization to demands from all stakeholders.

Innovation does not happen by innovators alone. We must be careful not to fall into a belief that innovation is reserved for one person or a special team whose primary function is to develop solutions to problems or invention to opportunities. Innovation is primarily cultural and thrives in team-based organizations. Avoid the trap that innovation is for a select few and all others are discounted. Innovation happens best when it becomes the culture of the entire organization and everyone has the opportunity to engage.

Working Cross-Functionally to Empower Patients to Self-Schedule Their Care

By Aaron Neinstein, MD, Vice President of Digital Health for UCSF Health, Associate Professor of Medicine, and Senior Director of the Center for Digital Health Innovation at the University of California, San Francisco

The COVID-19 pandemic was the catalyst for change in numerous facets of healthcare: The ways patients search for and find care, the ways health systems triage and manage care, and the ways doctors and clinicians provide care have all made enormous strides forward. At UCSF Health, we recognized early on that we could accomplish more, and faster, by working together, ultimately sowing the seeds of our digital transformation.

DOI: 10.4324/9781003372608-8

One of our earliest wins came within days of San Francisco's mandate that healthcare workers be screened for COVID-19 symptoms prior to entering the workplace. It was cold and wet that early spring, and hundreds of employees had to wait in long lines to enter the hospital as their shift started, often for as long as half an hour. The team at the Center for Digital Health Innovation (CDHI) at the University of California San Francisco (UCSF) quickly pivoted to develop an online symptom checker[1] that staff could use on their computer or smart phone before heading to work. Those with no symptoms received a "fast pass" that allowed them to bypass the screening line, and employees with symptoms were triaged to occupational health for further testing or care. This technology was later expanded for visitors to the hospital, including people making deliveries or visiting patients.

Similarly, our COVID-19 hotline was inundated with calls from concerned patients. With a flu symptom checker already in place through MyChart—a health portal for patients to access medical information and their care team—we were able to quickly roll out a version for COVID-19[2] that provided home care information or directed patients for testing and in-person care where necessary. It helped to reduce the burden on frontline staff and the hotline, and to manage and triage scheduling patients, thus increasing efficiency and cost savings.

As staff moved rapidly into remote work, other workflow issues impacted productivity and have led to lasting change. One such area was the need to virtualize the handling of incoming fax requests to process prior authorizations, durable medical equipment requests, and statements of medical necessity in a timely manner. These paper-based workflows typically were taking four days to process, but by leveraging an eSignature workflow through DocuSign—an application to electronically sign documents—we were able to reduce the turnaround time to less than one day and with most documents signed within hours.

A final example pertains to patients who received lung transplants at UCSF, who are scattered throughout California, the western United States, and beyond. Even prior to the pandemic, it was a challenge for patients to travel for regular follow-up care and testing and for providers to travel to regional centers for patients located further afield. COVID-19 presented enormous risk to these immunocompromised patients. This led to the development of a home spirometry kit that paired a device to test lung function at home with a virtual chat program.[3] This enabled doctors to be able to remotely monitor their immunocompromised patients for symptoms of infection or organ rejection and quickly intervene when necessary.

Successful digital solutions must take patient, staff, and provider experience into account; make financial sense to the institution; and ensure equitable access to care is not compromised. They must also make use of data to learn, iterate, and improve. None of this can happen in silos, and what the earlier examples all have in common are people from different teams coming together to collaborate and bring their area of expertise to the table, whether product management, human-centered design, clinical experience, or technical expertise.

Breaking Down Organizational Silos

In its historical state, UCSF had a fragmented patient experience where teams were not aligned toward significant, common goals—we lacked a holistic understanding and accountability for the end-to-end patient experience.

As part of our strategic mission to become a digital-first care delivery system, the organization has been undergoing a digital transformation to embrace a new way of working together while putting our customers—patients, families, caregivers, providers, and staff—at the forefront as we innovate at scale to shape the future of healthcare.

Traditionally, teams were siloed across the organization, working on their own projects with little to no crossover. For example, one team worked on patient feedback surveys while another team monitored star ratings and patient reviews on the web. These two teams were working *separately* toward the *same goal* of obtaining patient data, yet these processes were not streamlined and were inefficient and siloed.

By working cross-functionally, teams can break out of their historical silos and bring all the data together to best serve the needs of the consumer and of the organization. As UCSF Health Chief Marketing and Brand Experience Officer Sarah Sanders put it aptly, "To gain influence, lose control."

Now, each cross-functional team is composed of a diverse group of people and expertise with broad skill sets such as marketing, design, product management, solutions architecture, data analytics, project management, and operations. Each team is "self-sufficient" and possesses all that it needs to complete its projects. The diversity of the teams enables the team members to work together to find out what works, what doesn't, and why. These teams are also persistent, in that, unlike traditional teams, they do not disband after a project but are centered around a particular need like seeking care and will continually focus on experimenting, iterating, and improving based on feedback from their customers.

Empowering Patients with Web Self-Scheduling Appointments

This past year, we have begun to see the fruition of this new way of working by providing patients with the ability to self-schedule their own appointments from the web.

Web self-scheduling includes digital tools that aid schedulers to collect needed information and to offer patients their choice of time, date, and location for their appointments, as well as earlier appointment waitlisting and the ability to cancel or rebook appointments.

Patients across several of UCSF Health's specialty clinics are now empowered to advocate for their own health and can self-schedule their appointments online at their own convenience any day or time.

By making self-scheduling simple and straightforward, the tedious, time-consuming, and complicated processes patients face in healthcare are removed, speeding the path to the critical specialty care that UCSF Health provides.

For instance, new patients seeking care at our Cancer Center no longer have to endure long hold times, phone tag, or fill out endless forms. By answering a few quick questions online, they're connected with a patient navigator (as soon as the next day) who ensures they're in the right clinic, determines whether they need more tests, and matches them with the best provider appointment for their unique needs.

What the Data Shows

To better understand when patients are scheduling their Cancer Clinic appointments, we examined web traffic for scheduling activities on its website. The team found people were active at all times of the day and every day of the week—and much of this activity was outside of business hours. How many patients were previously having to defer scheduling appointments because our office hours were inconvenient?

By taking a closer look at web activity during different times of the day, we found that a greater percentage of users who were completely new to the ucsfhealth.org website were active online outside of our business hours and were scheduling appointments on their very first visit.

Going beyond the general web behavior into the actual days and times people are booking appointments, we found 38% of the appointments were booked outside of UCSF Health business hours, allowing patients to make appointments at their convenience. People made appointments throughout the day but most frequently close to the lunch hour, just before bedtime, and occasionally in the middle of the night.

We never would have been able to accurately glean these insights if a cross-functional team hadn't organized around this problem, and more importantly, patients would have lost valuable time in accessing care and may have chosen another hospital entirely.

Conclusions from Working as a Cross-Functional Team

As we continue our digital transformation journey at UCSF Health, we're continually learning and evolving to better our processes. Almost three years in, we've identified a number of key factors to successfully collaborate in a cross-functional way.

1. **Role clarity.** Role clarity is important when forming cross-functional teams. People were used to their own departmental, org-chart based, strict role definition, which meant they didn't know each other's expertise and had to adapt to a new way of working together. This sometimes meant a difference in their day-to-day responsibilities and accountabilities.
2. **Decision-making process.** People were used to a hierarchical way of making decisions. In the new model, decision-making is transitioned to teams, who are empowered and encouraged to make decisions based on data and strategic alignment.
3. **Communication.** It is important to ensure that people understand "what the rules are and what they aren't," overly communicate, and share both success stories and failures across the organization and with other cross-functional teams. This brings transparency to the work, reveals key areas for further alignment, and allows cross-functional teams to share and learn from each other. Also, remember to orient the team around specific customer journeys, and encourage questions to ensure they have all the information necessary prior to starting projects.
4. **Trust.** Earlier in the digital transformation process, people were unsure of the boundaries, nervous to cross the line, and unsure of one another's experience. Building trust takes time, but it cultivates an environment where people can be candid with each other.[4]
5. **Commitment.** Recognize that there will be challenges along the way, but stay committed to the bigger opportunity and vision.

6. **The Right Skills.** Ensure teams have the right skills and expertise needed to make progress independently. Provide a forum for cross-functional teams to meet regularly to share ideas, feedback, and progress reports so that everyone is in alignment and aware of each team's progress.

7. **Data**. Instrument the customer journey to collect data. Develop and centralize a comprehensive data set to generate insights. They can be used to illuminate the customer journey, operational improvements, service recovery, or new products.

Elevating the Power of Sound

Jaimie Clark & Hank Capps

It's an early spring evening in Atlanta, and residents of downtown buildings have finally opened their balcony doors after a long winter. The city sounds different tonight, however. Instead of engines rumbling out of business parking decks, students listening to the latest hits on their way back to lofts, and bus doors wheezing as they open and shut, claps and air horns are coming from the balconies to recognize healthcare staff changing shifts.

This is far from your regular spring night. Local news isn't reporting on rising temperatures, unlikely fashion trends for the season or if the Atlanta Braves will finally return to championship glory. The world is in the early days of an unprecedented, once-in-a-generation pandemic claiming hundreds of lives daily. COVID-19 is challenging researchers, running against the ticking clock to contain the spread of a fatal virus, and healthcare professionals, caring for an ever-increasing number of sick patients with limited information and resources, working long-hour shifts, and isolating themselves from loved ones for an undetermined number of weeks.

As we clap from our windows, we understand the heroic efforts of these individuals in scrubs starting their shifts or returning to nearby hotels for much-deserved rest. Yet, what's harder to remember is that the burnout they're experiencing isn't a new problem caused by the COVID-19 pandemic. While the current global situation brought the feelings of stress, fear, and loneliness to the forefront of our collective consciousness, they have been prevalent in healthcare for a long time.

A National Academy of Medicine study from 2019 states that burnout affected 35% to 45% of nurses and 40% to 54% of physicians. Fast forward to a survey conducted with more than 1,100 health workers between June and September of 2020, the height of the pandemic, and 93% reported feeling stress and 86% anxiety. Of these, 76% reported exhaustion and burnout (Mental Health America).

Caregivers are just tired and in need of care. The situation prompted our teams at Wellstar Health and Catalyst by Wellstar to seek innovative solutions for the problem. While there are vast amounts of audio-visual resources available online, including hundreds of apps and virtual reality options on every device, we wanted a solution that didn't introduce extra screen time to team members who already stare at monitors all day long.

It was also clear that such a project demanded a level of agility that is not always synonymous with healthcare organizations, especially when navigating through the uncharted waters of the first global pandemic in more than a century.

An Unlikely Partnership

Traditional health systems are large operations with decades of experience and large crews forming part of our communities' bedrock. Wellstar, one of the largest and most integrated healthcare systems in Georgia, is one of these organizations, a not-for-profit health system with more than 24,000 team members caring for 1.6 million patients every year in nine hospitals, five health parks, and more than 300 office locations.

However, much like a large battleship that requires absolute precision, these organizations can't always pivot quickly or adapt to new circumstances with speed. But imagine if these systems could launch speedboats to explore shallow waters, identify obstacles, and bring back solutions.

Enter Catalyst by Wellstar, the first-of-its-kind global digital health and innovation company, operated within a health system, to holistically address healthcare disruption by harnessing problems, solutions, investments, and partnerships across industries.

To disrupt healthcare and reimagine the future, Catalyst by Wellstar works in three phases.

First, we source needs and opportunities by listening to consumers, communities, and healthcare professionals to identify areas where we can make a dramatic impact. From these pain points and opportunities, we seek and partner with minds in the healthcare industry and far beyond. Lastly, after piloting successful solutions, we work with partners who can distribute them locally and globally, improving healthcare for everyone.

The work at Catalyst by Wellstar concentrates on six strategic focus areas that will shape healthcare over the next 10 years: (1) Digital health; (2) data, analytics, and security; (3) supply chain, logistics, and mobility; (4) customer experience; (5) sustainability; and (6) future of work. Each of these areas presents a wide net of opportunities to change the lives of patients and healthcare professionals, directly and indirectly. Together, innovation across these themes takes us closer to a future where healthcare is safer, widely accessible, more affordable, highly personalized, and attractive for a new generation of professionals.

The pandemic ushered in a wave of new ways to work for people across the globe, but for those jobs that cannot be performed away from the physical workplace, improvements to wellbeing did not come as quickly. For one of Catalyst's initial significant projects, we had the opportunity to help our healthcare professionals through one of the most challenging times of their careers.

Looking at the future of work strategic focus area, the recently formed Catalyst by Wellstar team put into action our belief that not all healthcare solutions need to come from our industry. Observing and partnering with companies across other sectors can inject fresh ideas and perspectives into healthcare and accelerate innovation. One of the first of these partnerships happened with Spatial, an award-winning immersive sound company known for their unique experiences in retail, hotels, museums, and offices. We asked whether we could create an experience together that would provide respite to weary caregivers with this screen-free audio experience.

But we still needed a space where these experiences could happen.

The Wellstar Foundation was partnering with Team Member Wellness to introduce 16 Wellness Rooms across Wellstar. Here, team members could unwind during their shifts in a safe and serene environment featuring dimming lights, fresh scents, and massaging chairs. It was the perfect opportunity to introduce sound.

"At our core, Wellstar is an organization focused on people, and that includes caring for the health and well-being of our more than 24,000 dedicated team members," said Julie Teer, senior vice president and Foundation president at Wellstar. "A healthy healthcare workforce is essential in our ability to serve the diverse needs of our patients and communities."

"While we can shut our eyes off, there's no way to close our ears," said Howard Rose, Head of Health and Wellness at Spatial. "Nurses are be subjected to alarms, over and over repeatedly, when there's nothing going on. So, what's happening is that their brains are going 'problem, no problem, problem, no problem.'"

For a healthcare professional, this always-on sense that registers a multitude of beeps and signals throughout a shift can take its toll in the form of stress and anxiety. In every Wellness Room, we can offer them a space where they can turn off for a few minutes and focus on their wellbeing. While we tested the implementation of virtual reality systems in these spaces, we decided that sound was a powerful way to relax the brain without adding further screen fatigue.

The rooms can be booked from any computer or smartphone, and there are no limits on how many visits a team member can make. Tablets outside each room also show the current schedule, and walk-ins are welcome if there are no bookings at that time.

When using the wellness rooms, team members can choose from several soundscapes, ranging from a deep-sea dive to a rainy day. These soundscapes also feature an extra version that has a low-frequency pulse to relax the nervous system, producing a calming effect on the body. Speakers strategically located throughout the room help create an immersive environment, as sounds can move closer or farther away, encircle you, or just cross the space from one side to the other.

Loud and Clear: What Team Members Are Saying and Where to Go Next

The Wellstar Wellness Rooms opened in January 2022, and team members across the organization have embraced the chance to reflect and recharge their batteries with a simple tap of their badge. Nearly 5,000 team members have used the rooms, with 95% reporting a positive mental impact. The self-reported mental state also increased by over 28%. During this time, 30,750 hours of soundscapes were played—the equivalent of 3.5 years—and 84% of users said they were highly likely to recommend the room to a colleague.

Each room also allows team members to leave their thoughts and feedback. Some of their quotes are captured here:

"A very calm environment that allows me to decompress after all of the negative energy that comes from seeing illness and injury throughout a day."

"Having this room is so helpful to the day-to-day stress we encounter not just at work but in life."

"A great way to relax after a stressful shift before I head home—allows me to decompress."

Catalyst by Wellstar gathered this real-world evidence first as a pilot with team members. We then expanded the healing power of sound to patients by installing immersive sound experiences in the labor and delivery rooms at one of Wellstar's hospitals. This new project is allowing our teams to test innovative ways to improve patient experience and wellbeing through this built environment.

The collaboration between Catalyst by Wellstar and Spatial highlights important points that can drive the future of healthcare across diverse systems of all sizes in all corners of the country and, eventually, the world. These are key takeaways that we must remind ourselves about as we embark on innovative journeys to shape the future of healthcare:

1. Team member wellbeing is paramount. Our clinical staff needs to be cared for to provide their best care for patients. At Wellstar, we like to say that every team member is caring for patients or caring for someone who cares for patients. Promoting mental and physical wellbeing means actively seeking solutions that protect, value, and reward each team member.

2. Launch that speedboat. Large battleships have enough inertia and resources to keep moving forward, even when engines stop running. They're well-oiled machines that have stood the test of time and the turbulent waters along the way. But now more than ever, it's necessary to unlock the power and agility of the speedboat, sending it out to explore unchartered waters, find new solutions, and report back on obstacles ahead. While the battleship provides what patients need today, your speedboat can seek what they want five or ten years from now.

3. The solutions that will revolutionize healthcare will not all come from the healthcare industry. It's essential to start looking at other sectors and invest in the agility and creativity of startups seeking to disrupt the way we work and serve others. We need to be encouraged by these disruptions and present them to stakeholders not as threats to what we do, but as opportunities to provide better care in an environment no longer limited by brick-and-mortar buildings.

By collaborating with multiple Wellstar teams and departments, as well as partners inside and outside the healthcare industry, Catalyst by Wellstar is charting new waters and inviting everyone to come along with us.

References

Mental Health America. 2020. *The Mental Health of Healthcare Workers in COVID-19*. https://mhanational.org/mental-health-healthcare-workers-covid-19.

National Academies of Sciences, Engineering, and Medicine. 2019. *Taking Action Against Clinician Burnout: A Systems Approach to Professional Well-Being*. Washington, DC: The National Academies Press. https://doi. org/10.17226/25521.

Appreciate the Complexity of Attention That Innovation Requires and Expose the Organization to Demands from All Stakeholders

By Leah Rosengaus, MSC, Director, Digital Health, Stanford Health Care; Michael A. Pfeffer, MD, FACP, Chief Information Officer and Associate Dean, Technology and Digital Solutions, Stanford Health Care and School of Medicine, Clinical Professor of Medicine, Stanford University School of Medicine; and Christopher (Topher) Sharp, MD, Chief Medical Information Officer, Stanford Health Care, Clinical Professor of Medicine, Stanford University School of Medicine

eConsults as Innovation

Tech-Enabled Relationships

Digital health service design challenges us to think differently about the clinical model, business model, and the enabling technologies we use to deliver care. In our view, innovation in this space almost always implies transformation of at least two, and often all three of these dimensions. This amount of change calls for cross-functional engagement, creative approaches, and a willingness to iterate.

The clinical model outlines "what, how, and who." It includes the actual clinical service provided, site or modality of care, and the service provider. The business model defines the target population, value proposition by stakeholder, cost, and revenue streams. Both are supported by enabling technologies that drive workflow and user experience. We will share our perspective on this three-pronged framework and our stumbles and lessons learned through the lens of our experience developing a novel eConsult program within Stanford Health Care and into the local community.

Setting the Stage

A general oncologist learns they share a medical scribe with a regional urologic oncologist. In a rushed moment between cases, the general oncologist writes up a clinical question on a scrap of paper, stuffs it into the scribe's pocket, and asks them to pass the note to their colleague across the bridge:

> *Text me Pz [xxx-xxx-xxxx]. 82yo urothelial CA resectable but not interested & no chemo interest "platin ineligible" plan Nivo "ready"? Call me/text:)*

To most, everything but the smile at the end of the note would be incomprehensible. But to a fellow physician, this classic "curbside consult" is part of the informal knowledge marketplace, conducted in clinical lingo, where clinicians barter on goodwill and relationships. Physicians and advanced practice providers (APPs) use these informal consults to confirm their thinking (e.g., "bounce" a differential diagnosis off a colleague), get quick answers (as the savvy community oncologist sought to do), continue their education, and even build comradery.[5] And so, for many reasons, curbside consults are a common and productive fixture of clinical practice.

Yet, there are pitfalls to the curbside:

1. *One needs to know who to call.* Curbsides rely on relationships a physician already has developed and are not conducive to forging new relationships. If you do not have a friendly link, you're out of luck.
2. *They can be a burden to specialists.* There is no direct incentive or compensation for the consultant's effort answering these questions.
3. *There is some risk.* The consultant has limited view into the relevant clinical data or history to give informed advice.
4. *The patient is a silent customer.* There is a missed opportunity to engage the patient in team-based care.

Enter the eConsult, an asynchronous provider to provider written consult, submitted by primary care providers or general specialists using structured clinical templates. The idea of eConsults originated as an approach to address specialty care access in the safety net, and it has since been adopted at health systems across the country, but adoption has shown wide variation in scale and value depending on how this solution is fit to the organization's capabilities and needs.

At Stanford Health Care, we adapted the model for both internal providers as well as external community partners. Despite the promising eConsult data out of peer health systems, we recognized a problem-first approach and a fresh data analysis were the critical starting points. Investing significant time and effort upfront to define the problem, size the opportunity, and develop the financial model were key to shoring up the value proposition specific to our own environment and not simply replicating a solution that had been successful elsewhere on hope alone. Through eConsults, we democratized the knowledge marketplace, leveraged technology to forge new relationships between providers, and started to hack away at access problems by empowering primary care providers to manage more cases.

Transforming the Clinical Model

Ambulatory care has changed little over generations, persisting a "one room, one patient, one doctor" model with limited opportunities for efficiencies of scale, team-based care, and quality driven by clinical collaboration. It further leads to disconnected care as specialty needs arise, worsened by limited specialty access that introduces further delays: The ability to answer a specialized clinical question and move on with the patient's care is limited. We sought to empower the primary physician, accelerate clinical pathways, improve the quality of decision-making, and enhance the prioritization of specialty care. We tested these outcomes with a pilot program in dermatology to show potential value. A comparison to traditional referrals indicated dramatic decreases in time to diagnosis and treatment plan of 23 days to 16 hours with the majority (73%) of eConsults being resolved electronically. In-person referrals from PhotoCareMD (27%) had a 50% lower cancellation rate compared with traditional referrals (11% versus 22%).

It was critical to know what the success criteria and ROI measures would be in advance of scaling and build the requisite infrastructure into the design to ensure we could measure it. We needed to address reporting capabilities from the start to ensure we could measure what mattered and understand whether we were solving the right problem. In addition to the metrics from our pilot program, we implemented a survey that asked the requesting provider, "What would you have done with this case if you did not have eConsults available?" Just under 60% of eConsults avoided a traditional referral, and 13% avoided a curbside consult. We also built in an eConsult dispositioning mechanism to understand how many cases would be definitively resolved via eConsult.

Over the first three years of the program, eConsults were expanded to 19 specialties and more than 120 diagnoses. Each diagnosis has an associated template designed together by a faculty specialist *and* a primary care doctor to guide the requisite clinical history, labs, imaging, or other data they need to provide a well-informed response on that issue. With the use of these collaborative templates, specialists were able to resolve 80% of eConsults without further follow-up.

Transforming the Business Model

If the curbside consult is the micro-interaction solving for clinical support, the macro issue at play is the lack of health system access to specialty care. At an academic medical center, faculty specialty expertise is perpetually a rate limiting factor for care. There is a fundamental supply-demand mismatch for specialists, manifesting in wait times from referral to appointment exceeding 50 days. An application of first principles thinking, "5 whys," or root cause analysis will point us back to multiple foundational problems, not all immediately actionable. However, the idea that our referral queues were inundated, in some cases with >20% lower complexity cases, was a demand-side issue we suspected we could better manage with referring providers and represented a compelling business case to the broader health system. The hypothesis tested in our MVP was that an eConsult program would help us to reshuffle the deck, getting lower complexity cases out of the queue while still ensuring those patients receive the care they need.

Working the business case from the consulting physician's perspective taught us that our specialists were not a monolith. There were some common themes, but they actually had fascinatingly distinct motivators. Some were driven by a desire to improve access and reduce wait times in their area, others had an intrinsic curiosity in digital health and innovative care models, and still others viewed this as a service to their colleagues aligned with their academic teaching mission. All appreciated the ability to be paid for otherwise uncompensated work. But there was an interesting phenomenon: Because the eConsult service is staffed by a small subset of specialists, it concentrates the consulting activity even more than the organic knowledge marketplace might have. This can shift the specialist's mindset from comparing the eConsult to previously uncompensated curbsides to comparing to clinic visits or other clinical activity they might perform. This points to the inextricable tie between clinical model, business model, and technology that must all evolve together. We cannot simply move a process from analog to digital without addressing its complexity. If we address technology but overlook the clinical and business model, we have (1) unintended consequences that we must then react to and (2) a missed opportunity to capture the full value of the service we could have reaped from an adoption, clinical quality, or business perspective if we had only harnessed the other two legs of the chair.

Enabling Technology

In the early days of product iteration, we recognized that the primary care providers initiate the eConsults and are therefore the linchpin to driving uptake. In product design, we make choices to get to a minimal *lovable* product rather than a minimal viable product (MVP) to promote the adoption of our primary care physicians. This meant the exchange added new value to them (unfettered access to the specialty knowledge marketplace), was readily accessible (imbedded in the EMR with synonyms and preference lists that prompted the eConsult order every time they considered a referral), and was delightfully easy to use (minimal clicks, no manual data entry and less than 2 minutes to submit).

In the curbside consult, patients were silent customers. With the new eConsult workflow, patients were not only aware of the interaction, but the specialists note was made visible to them in their chart so they could participate more actively in their care. But as with all MVPs, there are imperfections, such as the eConsult note being written for another provider versus plain language, and because the notes release rapidly to the online patient portal (similar to lab results), there is a chance the patient may see it before their primary care provider has an opportunity to engage them. Still, the response to the transparency has been overwhelmingly positive—as we've learned in other industries, customers want access to their data and to understand how it is being used.

Ultimately, our MVP demonstrated that the model would in fact build relationships between colleagues, empower primary care, and avoid unnecessary specialty referrals. We had our proof of concept and were ready to develop a scaled product for market.

Iterative Innovation Cycles to Prepare for Scale

Taking this product to market required an understanding of how our external customers were similar and different from our internal customers, what the market would bare, a clear understanding of fair market value and regulatory implications, and yes, a significant start-up investment. Starting with the customers in mind, we sought to understand what problems community providers and health systems were trying to solve.

Community providers are challenged on both sides of the coin. If they lack access to specialty care, it can be difficult to compete, and they may experience leakage to larger medical groups or systems. At the same time, these providers are often distributed, so even if they do have good referral relationships with specialty hubs, it may not be realistic for their patients to travel for ongoing care. Even more nuanced, the specialty access networks that community providers rely on look different everywhere. An eConsult offering that spans all specialties could be seen as competitive in a community that already has robust coverage in certain specialties.

Community providers also typically have much tighter IT and operational support resources, so a heavy product build or lots of integration may not be an option. Therefore, in external product design, we had to break our assumption that what we had built internally would work in the market. We started by testing a lightweight, EMR agnostic, technology platform and ensured we could "turn on" and "turn off" specialties based on the dynamics of the local market. The initial offering is more manual, but through a third-party clinical curation service, we were able to preserve the delightful ordering provider experience with no manual data entry. This approach gives both our organization and our partners an opportunity to test the clinical model and prove value before investing in significant software integration and IT infrastructure.

In Summary

Digital innovation that creates value at scale considers transformation of the clinical model, business model, and enabling technologies. Appreciating the complexity of the interplay between these pillars and how many unknowns we would encounter, an iterative approach to product development has served us well. Our friendly advice to colleagues developing novel digital health offerings and pulling these levers would be as follows:

1. *Take a problem-first approach.* Size the problem and opportunity using data that creates a strong business case in the organization's specific environment.

2. *Deeply understand your customers.* Appreciate the nuances of their needs, motivations, and perspectives. Do this rationally (hard value) and emotionally (what delights them).
3. *Iterate.* Leverage proof of concepts, pilots, and minimal lovable products to hedge the complexities of innovation and iterate.
4. *Scale smartly.* Recognize that taking the innovation to scale in the broader market requires a recycling of 1–3.
5. *Celebrate your wins with data and stories.* The success of innovation has a positive feedback cycle when you shine a light on it.

Through co-creating solutions, we enabled meaningful interactions between providers that benefited their patients supported by user-friendly, integrated technology solutions.

The Structure and Processes of Software Co-Creation at the Johns Hopkins Medicine Technology Innovation Center

Paul Nagy, Jasmine McNeil, Dwight Raum, Emily Marx, Amy Hushen & Stephanie Reel

The Johns Hopkins Technology Innovation Center (TIC) partners with care providers and clinical researchers to improve patient care through medical software. The TIC works with these change leaders to co-create novel health IT solutions and help scale those solutions to market. Launched in 2014, the TIC is a professional design and software development team serving as a central resource for the Johns Hopkins Medicine enterprise health system, including Johns Hopkins Hospital, home care, community and academic hospitals, as well as our insurance and accountable care organizations. In four years, the TIC has conducted development projects with 42 clinical leaders representing 35 clinical departments and organizations across our health system. Located on the main Johns Hopkins medical campus in Baltimore, the TIC is home to 29 developers, designers, project managers, and data architects. Through technical guidance and clinical partnerships, the TIC has helped launch nine companies scaling impact beyond Johns Hopkins Medicine into 69 other health systems.

Core Strengths

An internal, professional engineering team built inside a health system reduces the degrees of separation between developers that build solutions

and the care providers who use them. The TIC's on-site co-creation development process has advantages over traditional external software development when the solutions require intimate clinical domain knowledge, high
usability, dynamic clinical workflow, and changes in care providers' roles.
The TIC has expertise integrating clinical devices and information systems in
the development of clinical decision support tools and digital health patient
engagement applications. Clinical decision support tools aggregate health
records and generate predictive algorithms to help care providers diagnose
and treat patients with complex health conditions. Patient engagement tools
employ smartphones and wearables to enable convenient remote patient
monitoring and proactive care management.

The center embraces core principles of teamwork, respect, opportunity,
and audacity, with a pledge to stoke innovation in all corners of our institution. Nine values (Figure 8.1) define how the center curates and executes its
work. A few of those values are essential to defining the TIC's co-creation
culture when carrying out innovation.

Co-development: The TIC will only provide design and development support to projects—whether they come from clinical staff or Johns Hopkins
leadership—if there is a clinical champion who agrees to co-develop the
software with the TIC project team. This means that the clinical champion
must become part of the project team throughout design and development
sprints. The project team relies on the champion to provide expertise and
communicate needs based on their daily workflows. In turn, the technical
team uses the collaborative time to expose the clinical partner to the inner
workings of software development. Consequently, time to completion for
software projects is significantly lowered, feedback loops are faster, and the
clinical champion knows the details about how the project is progressing.

Patient-centered design: Other software development teams might use
designers at the end of a software project to work on the user interface. The
TIC practices patient-centered design throughout the design and development phases by leveraging its proximity to the hospital, clinicians, and their
patients. An integrated design team works throughout a project to identify
features and create a user experience based on the real needs of the direct
and indirect users. Developers, product managers, and clinicians are also
part of this design work, which includes one-on-one interviews, workflow
analysis, observation, design research, and facilitated group design sessions.
This focus on design that ultimately serves the patient also helps to expedite
development and ultimately creates software solutions that both clinicians
and their patients find intuitive.

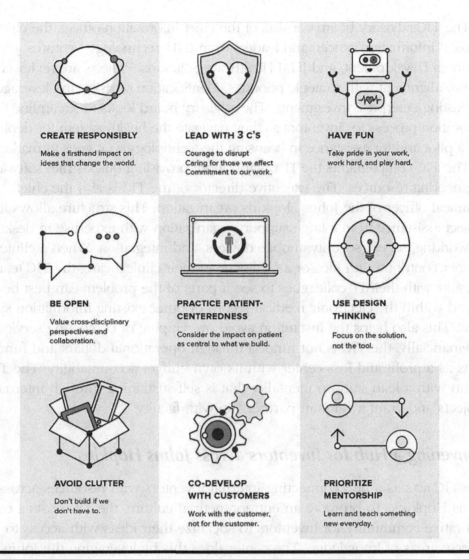

Figure 8.1 The core values of the Technology Innovation Center.

Aligning the Clinical, Technical, and Administrative Cultures within an Organization

Alignment across the clinical, administrative, and technical cultures is essential to exploring new technologies that can improve patient care in a way that is sustainable and feasible for care providers.

The TIC's location on the main medical campus with direct indoor access to the hospital increases the frequency of interaction with care providers. TIC staff visit clinical areas for requirements and usability studies that are facilitated by in-person observation and interview sessions.

The TIC advisory board consists of the chief information officer, the chief medical information officer, and leaders from JHU Technology Ventures, JHM Business Development, and JHM Healthcare Solutions. Projects are reviewed to ensure alignment with strategic priorities, identification of risks, and leverage to existing enterprise investments. The advisory board looks to streamline the innovation process for inventors as they navigate the health system for deploying a pilot and for assistance in licensing the technology as it goes to market.

The TIC complements the IT organization, providing quick-strike software engineering resources. The executive director of the TIC is also the chief technical officer of the Johns Hopkins organization. This structure allows for project assistance from a large support organization with expertise in desktop, networking, server, security, mobile devices, and integration. When a clinical partner contributes an idea or a challenge to their clinical care, the TIC team consults with their IT colleagues to see if parts of the problem can best be solved within the electronic medical record or other existing information systems. This also helps the institution avoid overlapping or competing services.

Financially, the TIC is not funded through operational dollars and functions as a profit and loss center with its own budget accountability. The TIC is run with a lean start-up mentality that is self-sustaining through internal projects and grant awards in partnership with faculty.

Convening a Hub for Inventors across Johns Hopkins

The TIC acts as a hub connecting internal inventors with resources across Johns Hopkins. To promote an entrepreneurial culture, the TIC hosts a collaborative community for inventors to socialize their ideas with access to diverse areas of knowledge. The center does this by leveraging the Johns Hopkins academic mission and by building active relationships with schools such as the Johns Hopkins Carey Business School, the Johns Hopkins Whiting School of Engineering, the Johns Hopkins Bloomberg School of Public Health, as well as the Maryland Institute College for Art. TIC staff regularly speak in these settings to recruit students and graduates and to enlist faculty to advise our teams. Beyond the academic community, the TIC draws upon the local business community to provide mentorship. The TIC uses its physical space as a natural base of operations for researchers, business, and industry partners to interact with clinical staff and Johns Hopkins engineering teams. The center's offices are built in a 10,000-square-foot open concept with abundant whiteboard walls and swing spaces for offsite partners to work and collaborate. The environment promotes impromptu design sessions, side-by-side developer/clinician collaboration, and introductions of cross-disciplinary subject-matter experts. The TIC connects future entrepreneurial team members in this hub space with clinical partners to understand a new project's scope.

TIC staff also host events and training programs across campus to bring the Johns Hopkins entrepreneurial community together. The largest annual event for the TIC is the annual Digital Health Day, which attracts clinical innovators from across Johns Hopkins to learn about and share innovation in digital health. In 2017, the TIC and its organization partners hosted 340 faculty and students along with 26 groups that research aspects of digital health and are piloting experiments in our clinical setting. Groups presented their work and shared experiences on digital health as a promising new opportunity for faculty career development, education, research, and improving patient care.

Building and Training Teams in Leadership

Hexcite is a 16-week adaptive leadership program that partners TIC clinicians with multidisciplinary teams to develop new ideas. Hexcite participants spend four months ideating and team building, with the following six months focused on building and deploying their pilot within the health system. The goal of this internal accelerator program is to enable rapid experimentation from ideation to deployment and evaluation within one year. Clinical leaders who apply must pay tuition and commit 20% of their time to join the program. Clinical participants split that time between cohort training sessions, team collaboration sessions, and interviews with potential customers. To enter the program, clinical leaders pitch a problem to a group of Johns Hopkins/local entrepreneurs and are evaluated on their leadership potential, idea, novelty, impact, feasibility, and organizational support. Participants are recruited from the following:

- Business students from the Johns Hopkins Carey Business School
- Engineering students from Johns Hopkins Whiting School of Engineering
- Design students from the Maryland Institute College of Art
- Design students from the Johns Hopkins Medicine Arts as Applied to Medicine program
- Other Johns Hopkins researchers and administrators

Teams participate in a series of workshops designed to develop early-stage start-up ideas into prototypes. The Technology Innovation Center uses its expertise in software design and development to help teams design a software product. Local start-ups and seasoned entrepreneurs serve as business teachers and mentors. Those entrepreneurs help teams to design their business using the evidence-based business model canvas.

By the end of the program, the teams have a practiced proposal for potential investors, mock-ups of their application, technical requirements (including architecture and data integration plans), an evaluation plan for testing their prototypes in a clinical space, a plan for obtaining seed funding (most often from translational grants), and a cohesive team ready to launch an independent start-up company. Hexcite primes teams to co-develop with the Technology Innovation Center once the 16 weeks are finished; the initial design and requirements planning allows the technical team to dive into development with fewer resources than a typical software development project.

What makes Hexcite unique among early-stage accelerators?

- It breaks down the barriers to co-creation that often inhibit innovation in large institutions.
- The TIC helps teams to navigate institutional security standards, Institutional Review Boards, data access and integration, marketing, technology transfer, and more.
- The program also seeks to build teams from the depth of expertise the institution offers instead of requiring that clinical leaders find resources on their own.
- Perhaps the most valuable element of the program is that the Technology Innovation Center staff train Hexcite teams in the IT principles of the institution so that they can easily integrate their technology and pilot it using an *internal* clinical founder.

The TIC continues to support the start-up teams that come out of Hexcite by providing technical mentorship, assisting with funding through grants support, connecting teams with other star-tup resources at the institution, and ultimately providing the technology team to build out the software solution.

The TIC identifies five key roles care providers bring to successfully co-creating solutions:

1. *Inventor.* They bring ideas and insights on how to fix broken healthcare systems.
2. *Learner.* A learner's mindset that aids the adoption of systems and design thinking to explore the potential solutions fully.
3. *Partner.* A peer partnership with the technical, design, and business team members to execute as a high-performance team.

4. *Scientist.* A scientific mindset to question their hypotheses and fully evaluate pilot solutions for learning what indeed improves patient outcomes.
5. *Leader.* They offer leadership in deploying new solutions and getting their peers to adopt new roles and technology.

Hexcite started in 2015 with yearly cohorts, and now after three cohorts, the TIC has helped teams gain more than $500,000 in start-up funding, trained 56 faculty/staff/students through the program, and helped 13 teams assess a combined 474 hypotheses about their technology and business. The training of technology and business leaders has led to intuitional gains beyond the start-up space. Participants reported the following learning improvements during the 2016 cohort:

■ Understood how to lead a technical project 36% ≥ 84%
■ Communicating the value of their clinical solutions 42% ≥ 89%
■ Effectively evaluating the competitive landscape 26% ≥ 89%

Decision Support Case Study: ReHAP

The process of co-creation that Hexcite brings to Johns Hopkins can best be understood by following one start-up team who participated in the first cohort of the program in 2015: ReHAP.

Dr. Krishnaj Gourab, a clinical partner and former chief of physical medicine and rehabilitation at a Johns Hopkins community hospital, was working to improve rehabilitation services to make sure patients who needed therapy the most weren't missed. Prior to Gourab's effort, therapists had no way to communicate which patients to prioritize between the multiple floors they needed to cover and often lost information about patient's therapy needs with shift changes.

With some mentorship from the Technology Innovation Center, Gourab built a low-fi prototype using MATLAB software to centralize the prioritization of patients and therapy caseload management. The software improved the therapy caseload management process for his team, and he quickly recognized a commercial opportunity as he interacted with rehabilitation groups from around the nation.

Gourab joined the Hexcite program to drive his commercial opportunity forward. He was paired with Johns Hopkins Home Care innovator John Adamovich, who brought financial and business expertise to the team. The two worked to complete the business model canvas through more than 40 interviews with potential customers. They also began to explore the data access and integration through workshops held with Johns Hopkins IT. Additionally, they learned about translational funding opportunities available to them, ultimately applying to and winning three separate grants to rebuild the technology as a secure application using the Johns Hopkins technology infrastructure.

According to Gourab, "The TIC helps us negotiate potential roadblocks when looking at the deployment so that ReHAP can work on the data."

Gourab and Adamovich were also able to refine their pitch through the program to gain further funding and leadership support for their project. They took these skills back to their roles at Johns Hopkins.

"This fellowship is unlike any other training I have had in my clinical career," said Gourab. He continued:

> *Apart from the training to create a healthcare technology start-up, it has exposed me to completely new ways of thinking while designing solutions in the healthcare arena. One of the most valuable parts of this fellowship is the focus on doing customer interviews and then to use the insights gained from the customer interviews to develop our products/ideas.*

The first prototype deployment of the application at the Johns Hopkins community hospital saved therapists an average of 20 minutes per day, freeing up more time for them to care for patients. Revenue increased after bed utilization in the acute inpatient rehabilitation unit went up from 36% to 72.5%.

Through the interviews that Gourab and Adamovich completed, they were able to secure five pilots at rehabilitation centers around the country. ReHAP is currently supporting those pilots and continuing to build its business.

Hexcite was developed to support and validate ideas like ReHAP through co-creation with business, technical, clinical, and design components.

Digital Health Case Study: EpiWatch

The success of co-creation around wearables and digital health is seen through a research study turned start-up working to alert patients with epilepsy and their caregivers before a seizure occurs.

EpiWatch, a research study and the first ResearchKit app built for Apple Watch, is the result of the collaboration between Dr. Gregory Krauss and Dr. Nathan Crone, both professors of neurology at Johns Hopkins Medicine. In 2015, Drs. Crone and Krauss began building the app from a research study into a seizure prediction tool to serve all epilepsy patients.

The app helps patients and caregivers manage epilepsy by tracking seizures, possible triggers, medications, and their side effects. Drs. Crone and Krauss were able to collect enough data through seizure tracking (with more than 1,000 study participants) to define a preliminary algorithm for seizure detection and alerts.

In order to make that data secure and accessible to researchers for algorithm development at Johns Hopkins, Drs. Crone and Krauss began working with the Technology Innovation Center in 2016. Backed by existing IT infrastructure at Johns Hopkins, the TIC developed a cloud-based architecture, allowing researchers to use a one-stop-shop for data collection, manipulation, and analysis with the ability to scale increasingly large data volumes from device sensors without requiring architectural modifications. The development team placed these tools in the hands of researchers to enable real-time comprehension and analysis of data. When a seizure event happens, the integrated system allows the researchers to immediately access, interpret, and export wearable data from that event. Leveraging this infrastructure has enabled refinement of the parameters that control the algorithm to help identify and reduce the number of inaccurate readings. The TIC also deployed design resources to interview patients and streamline app workflows (Figure 8.2).

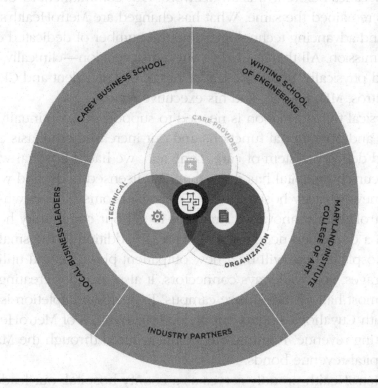

Figure 8.2   Aligning with the clinical, technical, and administrative cultures within an organization and connecting to resources outside the health system.

According to Dr. Crone: "It has really been a great experience working with the TIC because of the proximity and availability. . . . It is more of a collaborative relationship with people who have the right expertise for the project."

Here, the TIC's alignment with the institution's technical resources and its ability to pair that with the goals of two clinical change leaders in epilepsy research led to the first algorithm for seizure detection.

Scrappy Innovation in a Safety Net Hospital

Daniel Clark & Kyle Frantz

It takes evolution and innovation to thrive as a public hospital, especially in a city with a wealth of outstanding healthcare systems.

On May 6, 1837, Cleveland City Council designated an infirmary as City Hospital, a forerunner of the MetroHealth System. The initial charge was to care for the chronically ill, aged, mentally impaired, and poor—and 181 years later we are still doing that. Essentially, we have been practicing population health for decades before it was a buzzword. Our commitment to Cuyahoga County has remained the same. What has changed are MetroHealth's advanced and advancing technologies and the number of dedicated experts tied to our mission. All this has led to our transformation—clinically, operationally, and physically, under the leadership of our president and CEO, Akram Boutros, MD, FACHE, and his executive team.

The physical transformation is needed to support our continually improving clinical and operational functions and our increased emphasis on a value-based delivery system of care. Long ago, we had a hospital with 1,600 beds. Our current hospital has more than 700 licensed beds, and we are building a new 10-story hospital with 270 single rooms. This downsizing reflects MetroHealth's innovative strategies to get our community healthier, resulting in a decreased need for acute care. In addition to the smaller state-of-the-art hospital, there will be a new outpatient plaza, central utility plant, parking garages, and walkways/connectors. It also includes creating green space in almost half of the 52-acre campus. Expected completion is the end of 2022. With Cuyahoga County supplying less than 3% of MetroHealth's total operating revenue, funding was boldly secured through the May 2017 sale of hospital-revenue bonds.

While MetroHealth proudly serves as a county hospital, our leadership rejects the broad stereotypes that sometimes accompany safety net hospitals.

The senior leadership of the organization comes from a variety of backgrounds—public and private hospitals, consulting, technology leadership, private equity and venture capital, clinical and operational leadership, and academia among others. Our senior team and many of their direct reports have first-hand experience in organizations that are not structured like a public not-for-profit. We draw on that collective experience in everything we do. When cohort data is available, we are more likely to compare ourselves to our integrated delivery network or even for-profit counterparts than other public hospitals.

In a sea of health system consolidation, MetroHealth has remained unaffiliated. The growth has been steady over the decades and accelerated in recent years as clinical centers of excellence and nationally recognized physicians have joined our ranks. Our size—700+ beds at our primary academic medical center and $1 billion+ in operating revenue—works to our advantage when it comes to innovation. With 1.4 million patients a year, we have a meaningful cohort to implement new approaches to better health. At the same time, we are a size that allows for a strategy to quickly be understood and scaled across multiple facilities within what may be competing cultures and leaders.

In the last several years, we have begun to create an internal organization that is designed to support innovation. The Department of Integration and Transformation (DoIT) is the umbrella organization broadly charged with the transformation responsibility. The Center for DARE (Disruptive and Radical Experimentation) is a component of DoIT. DARE has a mission to "have a safe space to experiment with ideas" and challenge and "disrupt the status quo that is not working." DARE also acknowledges that everyone who works for the organization has an "individual passion" to advance how we provide care and treat patients, whether they are in a direct caregiver role or have a non-clinical focus.

The tenets of DARE are as follows:

■ We are scouts and scanners of our system, paying attention to emotional climate, morale, and commitment.
■ We work on culture, helping all of us have a voice in creating the culture that will help MetroHealth thrive.
■ We help "ideators" in the organization develop their ideas for change.
■ We "curate" information and resources about new ideas, best practices in innovation and leadership, and approaches to change.

Change management is also a critical component to making this culture change both initially and permanently. As we implement new technology and workflows, our leadership team believes that changing established processes is incrementally more difficult than changing technology alone. We often use the word "calcify" to describe how good processes corrupt easily and how those bad processes harden and take root. The goal is to break up the hardened calcium deposits so something new can emerge.

DARE also includes a focus on helping the organization change:

- We share best practices for employee involvement, working with resistance to change and communication approaches.
- We facilitate change and conversations.
- We convene a group to focus on critical issues.
- We bust silos and get people working in multidisciplinary groups to leverage a variety of perspectives and skills.

Like other organizations, project management and the Project Management Office serve more than the IT department. Recently, the PMO has been centralized outside IT and reports to a vice president who is responsible for both project management and analytics. These two functions serve the business needs of the organization, not just the reporting needs of IT. This separation of the reporting structure ensures that the data and information coming from our systems are the intellectual property and responsibility of the entire organization, not just the IT department. The co-creating enabled DARE structure has also helped give others in the organization a broader understanding of how projects are managed. Staff come to appreciate how complicated it is to manage a project—that it requires technical knowledge, attention to detail, political savvy, and an understanding of human behavior and motivation. In some instances, non-project managers have managed projects while being mentored by more experienced project managers. However, we also recognize the role of the professional project manager and believe having dedicated staff to work on critical initiatives—both big and small in the budget—contributes to the likelihood of a project being successful.

We embrace technology and were an early adopter of electronic health records. MetroHealth is Epic client #15 and first went live with computerized provider order entry in 1999. Over the decades, we have expanded our footprint considerably and now have one of the most well-integrated systems in the country. Our IS staff, approximately 140 team members, support a wide range of applications for the clinical and business organization. Operating at staff ratios well below national benchmarks and those recommended by our strategic technology suppliers, we manage a meaningful use that earns us national recognition.

Population health is a necessity when you serve a largely urban population. Early population health efforts were impressively successful. Through novel approaches, our Medicaid Waiver Demonstration Project generated claims that were nearly 30% lower than expected when compared with the CMS actuarial budget for the same patient population. To further manage this group, MetroHealth is a frontrunner for new models of accountable care organizations (ACOs) and has been chosen as one of 16 sites nationally to participate in a "Track 3" ACO model, which brings both upside and downside risk for the population attributed to MetroHealth.

All of this requires a combination of unique problem solving and analytics to support it. Dashboard solutions and pushing real-time comorbidity information to the electronic health record (EHR) is a combination of clinical care and technology that helps paint a more accurate picture of how sick our patients are. Moreover, recognizing social determinants of health makes evident the need for risk-adjusted reimbursement.

Outside of IS and the DoIT/DARE project management, a separate innovation structure keeps the organization attuned to new and developing technologies. Members of the team are encouraged to identify new technology that could be important to how we manage patients and organize the business. They then report to a senior vice president of innovation and strategy who is part of the CEO's cabinet. The innovation department follows by working with dedicated business development resources, IS, and DoIT/DARE to scope the opportunity, identify a budget and implementation plan, and take it through the governance process. We look for opportunities to identify and maximize alternative revenue streams inside and outside the health system while maximizing and leveraging our existing resources. We are seeking new technologies and partners as we drive healthcare innovation.

We see innovation as the key to everything we do and integral to fulfilling our mission and vision as the safety net, the county hospital in Northeast Ohio. We believe it so much that we include being "renowned for our innovation" as part of our organization's vision statement.

Mission

Leading the way to a healthier you and a healthier community through service, teaching, discovery, and teamwork.

Vision

MetroHealth will be the most admired public health system in the nation, renowned for our innovation, outcomes, service, and financial strength.

MetroHealth has used limited resources as a catalyst for innovation and community partnership throughout its history. MetroHealth, like most health systems, faces a constant challenge as to how to continually evolve the fundamental care delivery process and infrastructure (inclusive of technology and programmatic roles) to adequately care for our population in growing unmet social and clinical needs. This challenge is made more difficult by the pressure to deliver high-tech, high-touch, and highly effective chronic care management to both the general population and the attributable populations MetroHealth sees in the Cleveland area. Pursuant to these bold objectives, MetroHealth has launched more than 100 programs and initiatives designed to target various aspects of care for these patients with clinical and social determinants of health needs.

As a national pioneer for accepting new models of risk, the focus on finding value from the social determinants of health for MetroHealth could not be more pressing. Experts have found through repeated research that 50%–80% of outcomes and costs are directly impacted by the Institute of Medicine (IOM) categories of social determinants of health, inclusive of health literacy. The IOM made strong recommendations to HHS in 2016 to include social risk factor scoring for Medicare and Medicaid patients care strategies. Fundamentally, EHRs lack access to social determinants of health (SDOH) data and to the community resource inventories, while care plans and clinical decision support (CDS) predominately use clinical and claims data and are not firing based on social need, and thus CDS potential is not being realized. The continued shift from fee-for-service to value-based contracting is forcing providers/networks to become broadly responsible and accountable for patient outcomes. The data clearly demonstrates the link between the SDOH, care outcomes, and the overall cost of care. The cost of care is significantly higher for those patients who have ongoing unmet social needs.

MetroHealth has an almost exclusively employed provider population that is incentivized to collaborate. Combining this with the commitment to its community and willingness to assume the risk for patient outcomes, MetroHealth is an ideal early partner for Socially Determined—a

Washington, D.C.–based company that is creating the science of the social determinants of health. Together, our two groups are striving to quantify social risk and match it with targeted social interventions, while tracking the outcomes with traditional health system quality and outcome measures, thus forging the path for making social interventions an integrated and sustainable way of delivering care.

During the initial phase of the partnership, Socially Determined is developing a *SocialScape* for Cleveland and Cuyahoga County that quantifies social risk for the community at a census block level. Patterned after how an HCC score captures clinical complexity, the *SocialScape* captures the social complexity for economic well-being, housing, food insecurity, transportation, health literacy, social isolation, and crime and violence displayed geospatially, and each individual is assigned a risk score based on clinical and social factors. In parallel with this technical work, both entities are evaluating health system and community benefit programs that may best be leveraged for social interventions. Socially Determined is then using advanced analytics and machine learning to determine selected cohorts of patients who share similar social and medical traits that will benefit from selected interventions while outcomes are aligned with value-based contract incentives and tracked for efficacy. Once these findings are presented to MetroHealth leadership, a second phase of the partnership will begin with the execution and iteration of these interventions on other patient cohorts and will be funded by shared savings on value-based incentives and improved clinical outcomes. It is a partnership that both parties hope to share for the next three to five years and, while doing so, provide a roadmap for the future of U.S. healthcare while enhancing MetroHealth's role as a leader in innovation.

This relationship is one example of finding partners that are the right fit and targeting areas where we can make the most impact. The concept of non-clinical factors impacting health is not new to most in healthcare and is not new to MetroHealth. Collaborating with experts from outside allows us to leverage our assets and look at things differently, and it gives us the ability to develop a replicable solution that can impact more lives than we can physically touch today. Our care teams and employees have a tremendous amount of knowledge that, when supported by technology and innovation, can change the world of healthcare.

Targeting efforts that make our community stronger while helping other organizations get the most out of their limited resources is laying the foundation to implement future innovations. The willingness to adapt how and where we impact patients' lives is essential and fosters innovation. Limited resources helped make that adaptive excellence a part of who MetroHealth is. It has taught us not just to survive, but lead.

Innovations in Interventional Informatics

Christopher Longhurst & Marissa Ventura

As a practicing physician and the chief medical officer at UC San Diego Health, driving and supporting innovation in health IT across my team, organization, and beyond is a key part of what motivates me to do my part to improve the delivery of healthcare.

I have taken to heart the concept of the Triple Aim [1] to optimize health system performance and have embraced its expansion to the Quadruple Aim [2]. This approach encourages healthcare institutions to pursue improvements in the following:

- Population health
- Patient experience
- High-value care
- Provider experience

With this four-pronged approach in mind, I encourage my leadership team and my peers on the executive team to design health IT solutions that align with these goals—and meeting all four is the Holy Grail! To do that, co-creating solutions to the critical issues we face in our own organization, across our hospital and practice affiliates, and in the healthcare industry as a whole is a key facet of the work we do today and the work we have ahead of us.

In my own experience, bringing the right people together to collaborate on new ideas in health IT could not have been more important in most of the projects I've been involved with in my career.

Integrating Home Monitoring in the EHR at Stanford Children's Health

During my time as chief medical information officer at Stanford Children's Health, I helped lead strategic efforts to improve children's health and provider workflow using health IT. One way we did this was to innovate a way to integrate continuous glucose monitoring (CGM) data in the electronic health record (EHR) using widely available and reliable consumer technology [3].

Working in both pediatrics and clinical informatics, my colleagues and I recognized that Type 1 diabetes is one of the most common chronic diseases of childhood. Ensuring the appropriate level of insulin therapy could mean the difference between needing to manage hyperglycemia (high blood glucose) with aggressive therapy and inadvertently putting a young patient at risk for hypoglycemia (low blood glucose).

For years, we had recognized a need for a more straightforward way to collect and analyze home glucose data through what has become the center of our workflows: The EHR. While some cloud-based, self-monitoring diabetes applications exist, their use has been limited given the plethora of applications and EHR platforms.

A yet-to-be-tested solution to this problem began to take shape in 2014: Apple announced that it would update its operating system with HealthKit [4], a now well-known app that enables health data interoperability. Two additional announcements followed: Our EHR vendor (Epic Systems Corp., Madison, WI) and a major CGM device company (Dexcom, San Diego, CA) both announced they would leverage HealthKit for their patient-facing applications—the Epic MyChart app (patient portal) and Dexcom Share app, respectively.

By partnering with Dr. Rajiv Kumar, a pediatric endocrinologist at Lucile Packard Children's Hospital and junior faculty member at Stanford University, we brought together the technology platforms from these diverse industries: A medical device company, a mobile device company, and an EHR software provider. Getting these three platforms to "talk" to each other, we thought, could enable providers to better triage care between clinic visits and improve our workflow—features that would ultimately lead to more effective and efficient patient care.

While blending these platforms would help us determine the feasibility of integrating automated data into an EHR, we were also keenly aware that the process would be most useful in healthcare settings if the workflow could be replicated without the need for institutional-level customization or specific technology platforms. Moreover, to increase the likelihood of adoption among providers, we realized that there needed to be an easy way to visualize patients' glucose values over time. To overcome this potential barrier, my colleagues and I developed a model day visualization with a custom webservice embedded in the EHR. This clinical decision support tool is publicly available at https://gluvue.stanfordchildrens.org. It is agnostic to EHR vendor, and the workflow requires no institutional-level customization.

Our pilot demonstrated two things, quoting from our published findings in the *Journal of the American Medical Informatics Association*:

> *First, continuous information delivery is feasible through the use of commonly owned mobile devices. Second, passive EHR-based data delivery, coupled with automated triage and intuitive visualization, facilities more efficient provider workflow for reviewing data and improved communications with our patients. In our pilot, this was associated with better care between scheduled clinic visits [3].*

Fitting the pieces of the puzzle together enhanced the patient experience through improved care and improved the provider experience with a more streamlined workflow for the diabetes care team. It also supported improved population health with an EHR-integrated, risk-prioritized dashboard of all diabetes patients using CGM monitoring.

Bedside Tablets at UC San Diego Health

When I moved to UC San Diego Health as CIO, the 2016 opening of Jacobs Medical Center provided the opportunity for my new team to introduce what has become an often-cited, local patient satisfier: Connected iPads in every inpatient room [5].

Our tablets allow patients to be in charge of many aspects of their own inpatient experience with a simple swipe or tap. They have secure access to their own medical records through an app provided by our EHR vendor. Through this app, they can view their test results, photographs and biographies of their healthcare team, a schedule of medications or upcoming procedures, and educational materials prescribed by physicians.

Using Jamf for mobile device management, we are able to protect the patient's privacy upon discharge, as well. Each tablet is automatically wiped of a patient's data immediately following discharge, leaving no trace of the user's history. It is then ready for use by the next patient. Moreover, the software communicates with our Epic EHR system to coordinate iPad management with patient records.

The tablets also leverage a Crestron application to give patients control over their environment, including the blinds, room temperature (±3 degrees), and lighting. Entertainment apps are available with streaming to the 70" television in every patient room. Patients can also download some of their favorite apps. Our preliminary data suggest that this room automation is an attraction that is correlated with an increased likelihood of patients accessing their medical record.

The partnership with Apple was a nice carryover from my days at Stanford Children's Health. Our collaboration has helped us improve the patient experience at UC San Diego Health. Sometimes it is the simplest of amenities that help patients feel more in control and at home with their healthcare. The first week I saw patients in the Jacobs Medical Center, one new mother told me that, after her repeat Cesarean section, her husband had gone home to care for their first child, and having the tablet in the room helped her feel more in control.

A Hackathon at UC San Diego Health

When my team got involved with the hackathon at UC San Diego Health in 2017, little did we know the caliber of disruptive ideas that would come from bringing a multidisciplinary group of people together for a brief period to co-create solutions that address some of the critical issues in healthcare today.

Our first-place winner is a prime example: "Realty Art Therapy: Incentivizing Patient Mobility through Augmented Reality" was developed to help motivate patients to get out of bed and socialize with other patients as they search for art around Jacobs Medical Center. The app leveraged the iPads provided to patients during their hospital stay and took them on a hunt through the hospital's 150-piece therapeutic art collection. Each patient room includes an art piece, and the remaining pieces are scattered throughout the hallways and lobbies. The app is designed to promote patient mobility, diminish social isolation, and improve mental wellness through expressive arts therapy [6, 7].

With these potential health benefits, my team and I are helping the group of students who dreamed up the idea to develop the app further. One day it could become another tool in our arsenal to improve patients' experience and their well-being at our own hospitals.

Yet, what excites me about this project is the collaboration and partnership behind it. Our event, dubbed UC Health Hack, brought together technology professionals, engineers, clinicians, and undergraduate and graduate students to fulfill the ambitious goal of "closing healthcare gaps from the acute-care setting to precision medicine at home, empowering patients and their providers."

Nearly 200 students from institutions across California participated in the event. They were joined by more than 70 academic and industry experts who served as mentors, judges, or volunteers to help provide guidance and steer ideas into reality. Additionally, the event drew in notable partners, including UC Irvine Health, Rady Children's Hospital—San Diego, the UC San Diego student-led chapter of Engineering World Health, as well as some corporate sponsors, including Optimum, Amazon Web Services, Epic, Jamf, and West Health.

Partnering with Our UC Health Sisters

Collaboration and co-creation are at the heart of an important innovation for UC Health, the moniker for the University of California's five academic medical centers (AMCs) and 18 health professional schools. In a first-of-its-kind technology collaboration within UC Health, UC Irvine Health and UC San Diego Health began sharing the same instance of our EHR platform. This also marked the first time that our EHR platform had been extended from one AMC to another in the United States.

By sharing a platform with UC San Diego Health, UC Health cut the cost of implementation by an estimated 30%! This collaboration advances UC Health's strategic goal to share health services and generate efficiencies across campuses through the shared implementation and maintenance of technology platforms. It also promises to enable better management of medical information, help align clinical pathways and practices that leverage the best of both organizations, as well as better support joint research efforts.

Full implementation of the shared EHR platform with UC Irvine Health began in November 2017. UC San Diego Health also shares its EHR platform with the clinics at UC Riverside Health and community practice affiliates; this has been a cost-saving arrangement that has improved coordination of care among physicians [8, 9].

Again, the puzzle pieces came together in a way that has enabled us to address cost considerations, patient experience, and population health across a couple of our sister UC institutions in Southern California. When I cared for a newborn born to a surrogate mother at UC San Diego Health, the parents, who lived in Orange County, began asking how to best share the medical records with their pediatrician in that area. I was pleased to be able to share that my colleagues could easily access his child's medical records at UC Irvine Health because of our shared EHR platform.

Leveraging Apple's Health App—Again

Bringing us full circle, another collaboration with Apple is still in its infancy. In January 2018, Apple announced that the iOS 11.3 update would include a new feature to its Heath app: Patients would be able to view their medical records on their iPhones [10]. Among the dozen healthcare institutions invited as a pilot partner, the only state-run health system is UC San Diego Health [11].

This allows patients who have received care at UC San Diego Health or one of our affiliates, who are a part of Apple's beta program and have access to the iOS 11.3 (beta) release, to securely store their medical information from various institutions on their mobile phones. The data is encrypted and protected with the user's iPhone passcode.

The evolution of this new feature empowers patients with access to their own health records, consolidated in a secure place. Moreover, it has the potential to improve provider workflows as patients can share accurate information about their own health from a device in the palm of their hand. Will this make a difference to patients? Only time will tell, but we are committed to collaborating with patients, providers, and industry partners to continuously innovate.

Lessons Learned

As I review these projects, I noted that collaboration and partnership play a key role in helping to bring disruptive innovation forward. I also see common themes of the Quadruple Aim that we can pursue to drive and support innovation across health IT:

- How are you ensuring technology enhances the patient experience?
- How can health IT be expanded to improve the health of your community? Could the automation of glucose data in the EHR be used to analyze trends across young patients with diabetes? What lessons can we bring to our communities based on the data we aggregate in the EHR?
- How are you addressing the need to reduce the per capita cost of healthcare?
- What about healthcare providers? Let's develop health IT that allows healthcare providers to focus on what they do best—provide quality care.

One last thought: When you think of innovation, are you also thinking about the Quadruple Aim?

Notes

1 www.advisory.com/daily-briefing/2020/05/06/ucsf-health
2 www.ncbi.nlm.nih.gov/pmc/articles/PMC7184478/
3 www.centerfordigitalhealthinnovation.org/cdhi-portfolio-virtual-care-for-lung-transplant
4 www.centerfordigitalhealthinnovation.org/posts/the-secret-to-great-teamwork
5 www.ncbi.nlm.nih.gov/pmc/articles/PMC1435836/

References

1. Berwick DM, Nolan TW, Whittington J. The triple aim: care, health, and cost. Health Aff. 2008;27(3):759–769.
2. Bodenheimer T, Sinsky C. From triple to quadruple aim: care of the patient requires care of the provider. Ann Fam Med. 2014;12:573–576.
3. Kumar RB, Goren ND, Stark DE, Wall DP, Longhurst CA. Automated integration of continuous glucose monitor data in the electronic health record using consumer technology. J Am Med inform Assoc. 2016;23(3):532–537.
4. Apple. Apple announces iOS8 available September 17: Introduces new messages & photos features, quicktype keyboard, extensibility, iCloud drive & new health app. Cupertino, CA: Apple, 2014. www.apple.com/newsroom/2014/09/09Apple-Announces-iOS-8-Available-September-17/. Accessed for verification: March 2018.
5. UC San Diego Health. UC San Diego Health prioritizes patient experience with iPad and Apple TV: New hospital gives patients control of their environment. San Diego, CA: UC San Diego Health, 2016. https://health.ucsd.edu/news/releases/Pages/2016-12-06-uc-san-diego-prioritizes-patient-experience-with-ipads-and-apple-tv.aspx. Accessed for verification: March 2018.
6. UC San Diego Health. Students propose solutions to critical health issues at annual hackathon. San Diego, CA: UC San Diego Health, 2017. https://health.ucsd.edu/news/releases/pages/2017-03-23-uc-health-hackathon-solutions-to-health-issues.aspx. Accessed for verification: March 2018.
7. University of California. 2017 Sautter award winners announced. Oakland, CA: University of California, 2017. https://cio.ucop.edu/2017-sautter-award-winners-announced/. Accessed for verification: March 2018.
8. UC San Diego Health. Epic sharing with UC Health. San Diego, CA: UC San Diego Health, 2017. https://health.ucsd.edu/news/releases/pages/2017-11-14-epic-sharing-within-uc-health.aspx#sts=Campus collaborations allow for shared services, greater efficiencies. Accessed for verification: March 2018.
9. French R. Rollout of new electronic medical record partnership is an "Epic" achievement: Partnership between UCR and UCSD is first of its kind among UC Health organizations. *Inside UCR*. June 8, 2017. https://ucrtoday.ucr.edu/47454. Accessed for verification: March 2018.

10. Apple. Apple announces effortless solution bringing health records to iPhone: Health records brings together hospitals, clinics and the existing health app to give a fuller snapshot of health. Cupertino, CA: Apple, 2018. www.apple.com/newsroom/2018/01/apple-announces-effortless-solution-bringing-health-records-to-iPhone/. Accessed for verification: March 2018.
11. Sisson P. UCSD among 12 nationwide to pilot Apple Inc. new medical records system. *The San Diego Union-Tribune*. January 24, 2018. www.sandiegouniontribune.com/news/health/sd-no-apple-medicalrecords-20180124-story.html. Accessed for verification: March 2018.

Turning Everyday Employees into Start-Up-Like Entrepreneurs

Alex Goryachev

Cisco Fosters Culture of Innovation Company-Wide

"My Innovation" mantra encourages employees across all job functions and grades to tap into their inner entrepreneurs and start innovating in teams with a diversity of talents and skill.

Innovation Can Come from Anyone, Anywhere

> *Every industry in our digital age feels the pressure to move faster and find new ways to ignite internal innovation. Moreover, the goal is always the same: inspire employees to tap into their own passions and create game-changers for customers. The road map for this disruptive journey often takes different directions.*

I would like to share our method to unleash the inner entrepreneur of every employee and transform our workforce of more than 74,000 people worldwide into a start-up-like culture of constant innovation. We believe innovation can come from anyone, anywhere. Research and experience has validated this truth, especially when diverse and inclusive teams can access tools and resources.

No matter your industry or organization's size, our award-winning "My Innovation" program can help engage employees and inspire them to deliver brilliant new ideas—perhaps even the next big thing that disrupts markets with game-changing products, solutions, and services.

The Genesis of My Innovation

In early 2015, Cisco conducted a company-wide series of focus groups to get employees' ideas on how to improve the enterprise. One of the most striking themes from these focus groups was that employees wanted support for implementing innovations and a forum for sharing ideas that could shape the company's future. In response, Cisco launched a company-wide, cross-functional competition called the "Innovate Everywhere Challenge (IEC)" in September 2015. This competition provided a public, collaborative forum for employees to share, build, and invest in ideas for new products, services, and process improvements. By the end of the eight-month competition, 48% of employees across the company had participated in the Challenge. Of the 1,100 ideas submitted, seven remain active as new ventures. Moreover, the Challenge provided a space and process to discuss and build out new ideas and set the expectation that every employee can be an innovator.

The inaugural IEC became the catalyst for Cisco's company-wide support for entrepreneurship. Cisco brought all of these initiatives together under the umbrella of "My Innovation" and made this a centerpiece of the company's People Deal. To provide opportunities to build a start-up mindset and skills beyond the Challenge, Cisco created an all-in-one portal for all of its innovation work, assembled a network of mentors for new business ideas, added incentives for "Angels" who supported new ideas, and gave all employees the opportunity to make investment decisions in the second iteration of the IEC. The My Innovation framework enabled synergies among Cisco's many innovation supporters and led to a significant uptake in employee engagement in innovation. Now entering the third year of its sustained effort to empower every employee to act like an innovator, Cisco continues to learn more about how to embed a start-up culture in a large enterprise.

Tipping Point: The Decision to Invest in a Company-Wide Focus on Innovation

There is a widespread belief that large companies are unable to innovate at the speed needed to thrive in an era where start-ups can threaten established industries seemingly overnight. While many enterprises do struggle to adapt to new conditions, scale and innovation need not be at odds.

In fact, the diversity and sheer volume of employees at a large company can be an asset for sourcing new ideas if the company is able to harness and willing to act on them. The 2016 BCG Global Innovation Survey revealed that growth ideas most often come from internal sources and that the ability to tap into "employee ideation forums" was the greatest differentiator between strong and weak innovators. In other words, large companies who create the culture, space, structures, and incentives for their employees to innovate can be as effective at unleashing and harnessing new ideas as smaller companies.

In Spring 2015, Cisco held a series of focus groups across the company and discovered that employees were eager to innovate in their work but were unclear on the pathways for successfully nurturing new ideas to the execution phase. They asked for support, time, space, and money to foster and implement their best ideas for new ventures. In particular, employees wanted a central forum where they could bring new ideas, build them out with colleagues, and present them to executives and decision-makers.

Secure Executive and Peer Support

Cisco's Corporate Strategic Innovation Group (CSIG), together with partners across the company as well as external partners, began the work of mapping out how My Innovation could engage, empower, and enable *every* Cisco employee to innovate. Our first answer was the Innovate Everywhere Challenge (IEC).

During the planning phase of IEC, collaboration, research, and experience helped to garner full C-Suite commitment for the company-wide, cross-functional program. Additional collaboration with leaders across business units helped to map out goals for IEC:

- Capture disruptive venture ideas from employees and help grow them.
- Develop entrepreneurship skills and culture across Cisco.
- Enhance employee experience and collaboration.

MY INNOVATION MISSION

"My Innovation" empowers employees to innovate everywhere and at any time.

It is our employees' innovative ideas that will accelerate Cisco's leadership in the digital age. "My Innovation" creates the environment and provides vehicles for innovative ideas to emerge, develop, and shape Cisco's future. We disrupt the industry and ourselves by fostering a culture of grass-roots cross-functional collaboration, connecting our employees with a broader ecosystem of innovators inside and outside of the company and empowering everyone to take risks.

My Innovation, kicked off by the IEC, is now one of the pillars of Cisco's People Deal manifesto with employees, championed by CEO Chuck Robbins and sponsored by Chief Strategy Officer Hilton Romanski and Chief People Officer Fran Katsoudas.

Align with Corporate Priorities

Innovation has always been part of Cisco's DNA; however, we do not innovate for innovation's sake. Our goal is to create a disciplined yet flexible approach that leads to amazing business outcomes for customers and partners. To be successful, we needed structure and focus to make sure the IEC aligned with corporate priorities.

We incorporated flexibility and discipline into our guidelines. Flexibility that we would evaluate all entries. Discipline that all entries should focus on Cisco's strategic priorities and markets. For example, we published a Table of Strategic Innovation Elements, inspired by an external best practice, to guide our employees' ideation process. The table displays key markets, novel technologies, and disruptive business models.

Through our "Operations" market, we reinforced the need to develop innovations that reduce costs, streamline processes, or improve performance within Cisco or for our customers. To discourage "science experiments," we also emphasized "business outcomes" that are faster, leaner, and just better.

In addition, Cisco leaders across the company shared where they see a need for innovative and disruptive ideas—we call these Innovation Ambitions. With the commitment from the C-suite, employees were able to watch short videos that described each of the leader's ambitions (strategic priorities) that they could align their venture ideas.

Provide Clear Path to Innovation

The Innovate Everywhere Challenge urges all employees to Team Up. Disrupt. Innovate. Win. Just like in the external start-up world, IEC focuses on two personas central to creating a start-up culture:

- **Cisco Founders** create innovative ventures or join them at an early stage.
- **Cisco Angels** support innovative ventures financially or with expertise.

Each IEC proceeds over eight months in four phases modeled after the life cycle of start-ups: Ideate, validate, fund, and build. The IEC also includes a virtual investing component, giving employees the opportunity to participate in the selection process by investing free tokens (118,809 tokens invested in the last Challenge alone). These "backers" also posted 5,000+ comments and ratings, which provided invaluable feedback to IEC participants.

- **Ideate**: In the first phase, all employees are invited to propose a solution to a big problem or opportunity facing Cisco. They can create or join a cross-functional team to post their idea on a collaborative platform visible to all employees. A panel of judges—as well as employee token investment—narrow these entries down to 20 semifinalists, who advance to the Validate phase.
- **Validate**: The 20 semifinalist teams have three months to work with mentors and coaches to validate their venture, prepare an investor pitch, and fill a business model canvas. (See box on how they get help.) Cisco leaders with the help of real-world angel investors choose five of the six finalist teams; employees select the one finalist through token investing.
- **Fund**: The six finalists have three weeks to find internal sponsors and harness support for their venture. At the end of this phase, they give a 5-minute live pitch to a panel of Cisco C-Suite executives, external angel investors, and employees around the globe. The three winners of the Challenge are then announced at an all-company meeting and through various communication channels.
- **Build**: Each winning team gets $25,000 in seed funding, $25,000 as recognition, and the option to enter a three-month innovation rotation program to give them the space to develop their venture. Each team also receives a corporate concierge for help with practical affairs as well as assistance from Cisco leaders in determining the most appropriate next step for its venture.

Using a workshop format, Startup//Cisco equips employees with the skills and mindset of a start-up founder. Employees learn and apply innovation methodologies to accelerate new projects so they can get to stronger business outcomes faster and with fewer resources. Participants use design thinking and lean start-up innovation principles and techniques to validate ideas directly with customers. The process takes employees through designing a minimum viable service (MVS), gathering feedback from customers, analyzing lessons, revising the MVS, and repeating until Cisco and the customer agree that the MVS has been perfected.

Further, if semifinalists secure an executive sponsor at Cisco, IEC matches the sponsor's funds up to $10,000, hence encouraging business adoption. Finally, IEC also rewards employees who mentored, invested their tokens smartly (i.e., in semifinalist ventures), or more generally provided great value to ventures during the Challenge.

Excite, Inspire, and Engage the Workforce

Each IEC begins and ends with an interactive meeting broadcast company-wide featuring the CEO or other C-Suite executives who either kick it off or announce the winners. Throughout IEC, communications consistently inform and inspire employees through various multi-media, such as videos, articles, social media, presentations, and features on the progress of teams. Regular communications from corporate and business-unit leaders also reinforce key messages on the importance of internal innovation to Cisco's business strategy.

Finally, we make this serious business fun—we gamify it! Engagement levels soar when employees have creative ways to invest, root for their favorites online, hoist banners at events, search for like-minded team members, or brainstorm in online communities of their own making.

Empower and Equip Innovators All Year Round

The Innovate Everywhere Challenge is once a year. We needed something more to enable innovation year-round and constantly reinforce attitudes of innovation.

So we created "The Hub," an always on, go-to destination where employees come together to learn, collaborate, discover, and explore their innovative ideas. This interactive portal holds a wealth of resources to empower our Cisco Founders and Angels. Employees can develop their entrepreneurial skillset via playbooks and hands-on workshops, tap into a 4,000+ network to find the perfect mentor, get inspired through case studies of successes and failures, and more. The Hub is also home to IEC where our global employees can access and review the ventures any-where, anytime.

In addition to IEC, The Hub is currently the gateway to some 40 other innovation programs.

Impact

Innovation can and should be measured. We focus on five types of metrics: Engagement, participation, employee feedback, ventures outcomes, and thought leadership.

- **Engagement**: To date, 64% of Cisco employees, from all organizations and from 90 countries, have visited the Challenge website.
- **Participation**: 42% of Cisco employees have actively participated by submitting a venture, joining a team, investing their tokens, and commenting/rating.
- **Feedback**: 95% of employees surveyed would recommend IEC to their peers. Over the years, we learned from employee feedback and evolved the IEC process to what it is now.IEC gave me an opportunity to venture outside my comfort zone and interact with people I didn't usually interact with. After 32 years in the tech industry, the Innovate Everywhere Challenge helped me to discover that there was still a lot of innovation inside me.–

—Doris Singer

- **Outcomes**: The 45 top ventures from IEC1 and IEC2 have already directly yielded tangible innovations, including:
  - Proof of concepts
  - Beta products
  - 10+ patents in progress
  - Ventures adopted by the business
- **Example**: Cisco LifeChanger has been one of the most successful and disruptive solutions to come out of the Innovate Everywhere Challenge. It is helping to create new possibilities for how employers can tap into the tremendous untapped potential of people with disabilities, and change lives. What started as a novel idea by a passionate group of employees has evolved into a movement at Cisco, resulting in a best practice for other companies to emulate. Since its inception, Cisco LifeChanger has helped facilitate the hiring of nearly 100 people with disabilities globally, across all disabilities, into a broad range of functions, including sales, engineering, technical services, employee services, and customer advanced services, in addition to our general business functions.

- ■ **Recognition and awards**:
    - ■ Cisco LifeChanger has been recognized globally by industry-leading disability inclusion organizations, including: Disability Matters 2016 North America, and Disability Matters 2017 Asia Pacific.
    - ■ Cisco was a recipient of the 2017 Diversity Journal Innovation Award, and the USBLN also recognized Cisco as one of the "Best Places to Work for Disability Inclusion."
    - ■ To date, Cisco LifeChanger has also been featured as a disability employment best practice at national and global events, including the Zero Project Summit, Disability Matters, USBLN Global Employment Summit, International Labour Organization Global Disability Summit, and the Harkin Institute Global Disability Summit.
- ■ **Thought leadership**: Constantly reinforcing our innovator brand is key to attract and retain talent. Since its start three years ago, the Innovate Everywhere Challenge has contributed to Cisco gaining media recognition as a "Great place to work" as well as acknowledgment from Human Resource organizations as an industry leader for our "Employee Engagement" and "Unique or Innovative Talent Management Program."

The Next Frontier: My Innovation and IEC in Year Three

In Summer 2017, the Cisco team took stock of what had been achieved since the integrated My Innovation supporters came together under The Hub a year earlier. During fiscal year 2017, 53% of Cisco's employees (more than 39,000 people) engaged in My Innovation, and 23,000 of these participated actively. They came from 89 countries and all were part of Cisco organizations. This reach suggests that Cisco's innovation efforts have achieved the scale necessary for true cultural change.

Rather than resting on its laurels, Cisco has already identified two ways to strengthen My Innovation as it enters its third year. The first is supporting business owners in more clearly articulating the discrete design challenges that innovation should address. This will lead to better solutions both in the third iteration of the Innovate Everywhere Challenge as well as in Cisco's day-to-day business.

Cisco is also expanding its network of makerspaces through its thingQbator initiative. These physical spaces are dedicated to supporting hands-on learning, exploration, and prototyping and are themselves examples of how deeply the entrepreneurial mindset has permeated Cisco: The initiative started with a group of employees in Cisco's Bangalore office, and My Innovation is now helping scale it in the United States.

Conclusion

In 2015, Cisco committed to ensuring that it thinks and acts like a lean start-up but scales as an enterprise. The My Innovation journey has already directly yielded tangible innovations, including seven ventures with a successful proof of concept, two beta products, and eight patents in progress. Over time, these numbers will only grow.

Accomplishments of IEC1 and IEC2 Winners (as of November 2017)

Just as important for Cisco's future success has been the change in the company's culture: Employees now have the opportunity and support to propose new products, services, and process improvements and see themselves as innovators and entrepreneurs. As this expectation takes root among all employees, My Innovation will continue to evolve to ensure that everyone has the skill sets, attitudes, and opportunities to innovate.

With the Innovate Everywhere Challenge, "My Innovation" started a lasting cultural transformation. My Innovation will continue to evolve to ensure that everyone has the skill sets, attitudes, and opportunities to innovate as we launch the Innovate Everywhere Challenge 7.

Chapter 9

Conclusion

What follows is less of a summary and more of an admonishment. My hope is that this is not just another book with some good stories but a catalyst for you and me. A catalyst for our industry. A catalyst to push harder and more thoughtfully along this healthcare technology journey. The beauty of the innovation pathway is its simplicity. The processes are proven, and the framework widely adopted. While I do not believe there is a formula for innovation, this framework provides a solid process and pathway from which to innovate.

Blending cultures is a key place to start. You must make sure you have organizational buy-in and commitment. In cases where this is not possible, start small, gain experience, establish credibility and doors may open. I admit that occasionally, you just need to create a skunk's works of sorts and demonstrate value and ask for forgiveness later.

While it is critical to leverage technology, never start there. We are in the people business. You have to strike that perfect balance in your organization between people, process and technology. Win the confidence and heart of people, and everything else will fall into place.

Some may think planning is an innovation inhibitor, but our contributors have shown otherwise. Innovation success is often multiplied when you use sound planning principles and create roadmaps. Sometimes innovation just happens as needs arise, but roadmaps have a stronger track record.

John Maxwell says that "one is too small a number for greatness." I agree. Very few innovations come from a single person. They come from a team of teams or an agile environment where collaboration and communication dominate the culture. I do not recall any innovation that I have been involved with that emanated from just one person.

DOI: 10.4324/9781003372608-9

One of the key tasks of a leader is to eliminate barriers. Some of the contributors highlighted roadblocks to innovation and how they overcame them. Whenever you pioneer, you naturally encounter barriers. Accept this fact, and do not be discouraged. Rather, see barriers as signs you are on the right path. Then relentlessly eliminate them.

A primary key to Apple's success has been the diligent pursuit of simplicity. Too often we complicate matters, which in turn reduces our opportunity for success. Always seek to simplify the complex. The more focus, the sharper the solution. Innovation dies with complexity.

It's human nature to be driven by recognition and reward. We have a propensity to repeat those things rewarded. Create programs and incentives for innovation. It will eventually become the fabric of you and your organization. Celebrate failure.

Just as you seek to collaborate within your organization, consider including other key partners. Many of the examples in this book were instances of supplier and provider collaboration. Provider and patient collaboration is powerful as well. Often, a collaborator will have the missing piece for you to complete or enhance your innovation.

Now that you have completely read the book, I want to leave you with what I believe is the single biggest key for successful innovation: You. To be innovative, *you* must be innovative. You can take the innovation pathways framework and adopt it in your organization. As the stories demonstrate, when the framework is adopted well, it will work. If you want to get to the next level, work on yourself. Seek to be an innovative person if you are not already. Too often, I speak with people who are frustrated because of a lack of innovation in their organization. The first question I ask them is how do they make sure they are personally innovative? If they are still using kiosks at the airport, probably not innovative. If they still have a printer in their office, probably not innovative. If they are still going to their local bank or grocery store, probably not innovative. If they have the same hobbies as they did ten years ago, the same clothing and music styles . . . probably not innovative. If they don't routinely read and study inside and outside of their expertise, probably not innovative. Same phone forever? Same glasses? Same drink? Basically, if you are limiting your experiences, you are unable to take advantage of all the diversity in science, nature, tech, arts and philosophy. It is the combinations of these inputs that make one innovative.

So if you are not innovative, you will struggle to innovate.

The solution is simple and pragmatic. Do new things. Constantly introduce change into your life. Consume diverse news, and listen to contrarian viewpoints. Activate Twitter, Snapchat and Instagram. Download new music. Learn a new dance or hobby. Go back to school. Use all the tech in your car. Grab a mentor who is a generation or two younger. If you are a technologist, learn art and vice versa. I guarantee that as you become innovative, everything else will follow. And we will change the world.

My hope and prayer are that we will never experience such a devastating pandemic as we did with COVID. Through learning and innovation, may we be better prepared the next time.

Index